TIMELESS DOMAIN
Foundation of All Phenomena

Kien T Mai, MD

IBSN (book) 978-1-0679213-1-6

e book 978-1-0679213-0-9

TABLE OF CONTENTS

PREFACE

COULD TIMELESS DOMAIN BE THE FOUNDATION OF ALL PHENOMENA?

For centuries, philosophy, religion and science have sought answers to fundamental questions: Where does reality come from? What is time? How does consciousness arise? Why do matter, gravity, information, and mathematical laws exist? When a problem reaches the limits of knowledge, it often appears paradoxical. Examples include quantum entanglement between widely separated photons. When photons belong to the same system, no matter how far apart they are, they always manifest a perpetual connection; measuring the state of one photon instantly reveals the measurement of the other. In the double-slit experiment, without path detection, photons display wave-like behavior; when measured, they display particle-like behavior. This phenomenon reveals the profound role of pure observation. It highlights a deeply counterintuitive reality: the mere act of measurement can fundamentally alter quantum behavior, forcing a quantum entity to shift its manifestation from a wave into a definitive particle.

Paradoxes are numerous in life. They may reflect our failure to perceive a deeper foundation beneath observable phenomena. To an ancient society unfamiliar with electricity, even a glowing bulb suspended upside down could seem inexplicable in the 4D world

The Physics of Timelessness

Delving into the concept of Timeless Domain (TLD) a simple observation begins appearing: space and time need not share the same origin. Although modern physics unifies them as spacetime, they may not have emerged simultaneously. **Stephen Hawking** showed that near the Big Bang and in black hole physics, the classical concept of time no longer retains its familiar meaning. In his formulation of **imaginary time** and the **no-boundary proposal**, time may be treated mathematically in a manner similar to a spatial dimension, thereby blurring the conventional notion of what occurred "before the Big Bang." **Julian Barbour** went even further by proposing that time itself is an illusion. According to Barbour, the deeper reality may consist of timeless configurations, while our perception of the passage of time arises from the "records" or "time capsules" embedded within these configurations. **Carlo Rovelli**, through the framework of **Loop Quantum Gravity** and relational physics, likewise argues that time is not a fundamental entity existing independently of physical systems. Rather, time emerges from the relationships, changes, and interactions among physical systems. In this view, there is no absolute cosmic clock—only the evolving relationships between different parts of the universe. The **Wheeler–DeWitt equation**, often regarded as one of the fundamental equations of quantum cosmology, contains no explicit time variable in the conventional sense. The well-known **Wheeler–DeWitt equation**, often regarded as one of the fundamental equations of quantum cosmology, contains no explicit time variable in the conventional sense when the equation is applied to a constraint area in space $\hat{H}(x)|\Psi\rangle=0$

A simple analogy for TLD is a blank sheet of paper or an empty television screen. Although both are physical, their relevant feature is their capacity to contain and display information. An image may appear on a screen as a complete whole rather than being drawn line by line. Likewise, informational structures in TLD exist as complete possibilities and become manifest when appropriate conditions arise. Paris and New York may coexist on a map without requiring a temporal journey between them.

The Phenomenology of Dreams

During sleep, particularly during rapid eye movement (REM) sleep, when vivid dreaming commonly occurs and skeletal muscle activity is largely suppressed, a person may see scenes, hear voices, encounter other people, and experience narratives that seem to unfold over long periods. A person may be in one city and then instantly appear in another country without experiencing any journey between them. Yet these images and sounds do not arise from the actual propagation of external light or the physical oscillation of sound waves through the surrounding space. In this sense, dreams illustrate an intriguing possibility: **information can be experienced without necessarily being accompanied by a clear perception of physical time and space**.

Dreams, of course, remain products of brain activity and occur within biological and physical time. They provide a useful analogy of timeless domain concept. They demonstrate that human experience does not always depend upon the conscious perception of matter continuously moving through physical space. To date, the exact nature of time within dreams remains largely unstudied. However, the apparent non-existence or suspension of time during standard dreaming—such as in Rapid eye movement (REM) sleep—is built on the neurological evidence that muscle atonia (paralysis) prevents physical movement that always and requires time. REM sleep behavior disorder (RBD) when a person physically acts out because the muscle atonia during REM sleep fails. Conversely, in lucid dreaming, time perception is studied as a complete, continuous entity. Furthermore, in rare instances of NREM dreaming where muscle paralysis does not occur, the experience of certain type of time is present. All of this indicates that the role of time is not merely governed by static laws; rather, time can manifest in layered hierarchies—occurring as non-continuous (fragmented), continuous (total), or completely conditional upon the operational state.

From Emptiness to Symmetry Breaking with the Emergence of Spacetime: The Mechanics of Cosmological Phase Transition TLD

If time is emergent, a TLD must precede it since time needs a space to be recognized. TLD may therefore fill the missing conceptual gap between the Emptiness and spacetime in our current understanding.

Emptiness→ **TLD** →4D spacetime

Emptiness is not defined as nothingness but the absolute symmetry. As a result, the emergence of TLD is natural and inevitable. Without time, there can be no motion; therefore, velocity and acceleration lose their meaning. TLD is thus a Protospace. It contains no past, present, or future physical energy, mass, particles and consciousness. Information is non-temporally co-present, because there is no temporal flow TLD contains no physical matter, formal knowledge, energy, photons, or sound. Yet, it possesses infinite omniscience, knowing, and boundless energy held in a state of perfect supersymmetry and equilibrium. Within this realm, there is no sequential time, no birth or death, and no causal arising of one separate thing from another; instead, there is a seamless continuity of structure., LTD resembles Eastern notions of timelessness: unborn, undying, and beyond temporal change.

TLD contains Information. Unlike the primordial Emptiness, where absolute symmetry prevents manifestation, TLD is the mode of expression of Emptiness, a domain in which freedom permits the first degree of asymmetry to arise. Yet it contains no matter. It is therefore the only possibility that structural information represents its content.

This information may be integral from primordial balance, virtually limitless in its properties, protoforms, quantity, and structural mechanisms. It comprises the informational precursors of later structures. For example, principles eventually manifested physically as DNA structural information may already exist in TLD. All coexist without temporal succession.

Opposites such as strong and weak, positive and negative, may coexist without excluding one another, because they have not yet entered the dualistic world of spacetime. Moral conflicts are also absent, since such moral categories require time, action, and causality. All informational units coexist without before or after. They neither arise nor cease, nor must one produce another. Though distinct, they need not be linked by linear causality. Collectively, they may be conceived as Prajñā Wisdom, Ultimate

Omniscience, or Primordial Knowing—not the knowledge of an individual, but the inherent capacity for knowing within the foundation of reality itself.
Because (configurational) freedom naturally exists, it permits informational deviation, TLD may lose its nearly perfect symmetry. This creates a need to restore order and correct deviation, requiring a new mechanism arising from TLD itself: **time**. When temporalization arises, such possibilities provide the basis for stratification and ultimately triggering the phase transition that collapses potentiality into the localized, thermodynamic flow of our four-dimensional spacetime universe by process outlined later in the next paragraph.

When deviation is minimal, only ordering is required; physical transformation has not yet begun. This preliminary ordering mechanism is called proto-time t_0. As deviation grows capable of altering the state of TLD, information may be represented as pregeometric proto-filaments: integral information remains straight. The deviated information in TLD is characterized abstract informational variables that later acquire geometric meaning by inclination angle θ, curvature κ, and torsion τ. These changes constitute the three basic mode of natural geometrical change in the process of **pregeometrization**.

Biased information as the precondition for matter
When informational inclination exceeds a critical threshold, the natural need for correction and restoration becomes necessary so that the deviated informational structure can continue to be adjusted and perfected within TLD. The first condition of this corrective process is that the deviation must become manifest and capable of being recognized. For this to occur, biased information must undergo a process of increasingly explicit manifestation and ultimately become **materialized**—that is, actualized in the form of matter.
In the model proposed here, the tendency toward restoration is regarded as an intrinsic characteristic of the order of TLD. A deviation cannot be corrected if it remains completely latent; first,

it must be made sufficiently explicit so that the nature and consequences of that deviation itself can be recognized.

Therefore, the model proposes two levels of manifestation through which biased information becomes increasingly explicit and capable of being corrected. Time emerges as an imprint of this process of making deviation manifest. In this sense, time initially is not the cause of transformation, but rather the sign of a deeper process in which biased information is made manifest, ordered, and becomes an object capable of being recognized for correction.

i. Slight deviation—or information that cannot yet undergo complete correction likely producing measurable but non visible dark matter (DM) whose effective mass reflects informational curvature.

The time required for information to materialize, alongside the necessity of a proto-time t_0 to organize information with minimal deviation, suggests that DM contains an intermediate temporal component, provisionally designated t_1.

ii. Beyond a critical threshold, the degree of temporalization depends on the condition of the information. Time $\mathbf{t_2}$ produces visible, measurable matter.

This principle can be summarized as follows as the process of Information Pre-geometrization and Temporalization into Matter (IPTM):

Informational Deviation→ Need for Manifestation (Pre-geometrization) → Temporalization→ Deviation Becomes Manifest→ Correction and Restoration

Such information diverges from the canonical information due to information deviation causing fuzziness or obscuring the Ultimate Omniscience. This fuzziness can be mathematically represented by the following equation, which sums the squares of κ and τ (Because κ and τ are not collinear, their total value

mathematically suggests the sum of their squares, invoking a higher-dimensional variant of the Pythagorean theorem.

$$\begin{pmatrix} g\tau\tau & g\tau\kappa \\ g\kappa\tau & g\kappa\kappa \end{pmatrix} \rightarrow \text{Symmetry breaking} \rightarrow \text{invariant structural Energy}$$

$(\tau^2+\kappa^2)$

Matrix represents "section" of TLD that is symmetric, $g_{\tau\tau}$ describes the geometric fabric's measurement along the torsion/twist τ (tau) dimension, $g_{\kappa\kappa}$: the measurement along the curvature κ (kappa) dimension $g_{\tau\kappa}$:the interaction or cross-terms between twist and curvature. Note that θ is represented in κτ. In a matrix, the geometric **TLD** is defined before the interaction with time.

Symmetry breaking occurs because τ, κ are non-collinear (independent). Therefore, any microscopic fluctuation or "freedom" within the matrix causes a tiny mathematical imbalance. The system cannot remain perfectly symmetric forever; the configuration space becomes unstable, much like a ball sitting perfectly on top of a Mexican hat.

The introduction of time is to relieve this geometric instability. It collapses from its high-energy, timeless symmetric state into a lower-symmetry state.

This non-collinear relationship yields profound implications for how the TLD acts as the foundation for all observable phenomena:

Thus, within the IPTM hypothesis, τ could encode features such as chirality or handedness, twisting and helical organization, orientation and internal rotation, asymmetry, how localized matter is arranged into a particular form, **κ** is proposed to generate gravity as well as the curvature of space and matter,. However, morphology would arise from the combined state. the informational curvature κ and τ along with other parameter θ,ρ,ℓ,fields and interactions required to weave a complex, multi-dimensional topography out of absolute nothingness. This proves that complex, highly ordered information can exist statically within the TLD long before macroscopic time ever emerges.

M form=F(θ,κ,τ, ρ, ℓ,fields and interactions)

ρ- density of information, ℓ "length of infornation related to the size

Based on these parameters, one can derive the fundamental mathematical equation for the frequency ω and amplitude A of the vibration. From this information, the geometric configuration of the string after temporalization transforms it into a vibrating string within the interface—the intermediate realm between11-Dimensional Superstring Theory (11D) and Loop Quantum Gravity (LQG). This TLD with pre-geometrized information which always reside in a pre-space is orthogonal to the strings in 11D theory and LQG of the **4D** spacetime."

Physical time $\mathbf{t_2}$ emerges to geometrize and materialize deviated information into (ordinary) matter (OM). TLD does not literally become 4D spacetime; rather, it remains a deeper domain orthogonal to 4D—and, hypothetically, to the full 11-dimensional framework of M-theory. Integral information may be pictured as fibers extending orthogonally to spacetime, collectively forming an informational fiber bundle in TLD. Only deviated information undergoes geometrization and temporalization.

At this point, the TLD model may be compared—speculatively—with string theory. The three parameters θ,κ,τ act as primordial structural codes represents deviation, curvature associated with mass and gravity, and particle spin respectively. Through temporalization, pregeometric informational filaments become vibrating physical strings. Their purpose is stratification, manifestation, and eventual correction.

Photons first arise at the threshold of visible manifestation of information at the primordial angle θ, characterized as uncurved yet twisted. The energy of the photon originates from the TLD, providing just enough energy to initiate wave creation and achieve the speed of light c. Meanwhile, the neutrino is the lightest quantum particle after the photon in the current standard physics, corresponding to information that is slanted, gently curved, and twisted with a spin of ½. Hypothetical graviton: θ minimally exceeds the threshold, with no curvature κ=0, and torsion, spin =2, as manifestation of the gravity field. Lastly, the Higgs boson of the Higgs field manifests mass, which corresponds to information that is slanted and curved but untwisted to be served as model for mass configuration. In this model, curved and twisted compact geometries due to curved and twisted information are represented by matter exhibiting different aspects of Calabi–Yau spaces commonly seen in the 4D space-time.

Because matter cannot remain in timeless TLD, it is projected into 4D spacetime. As reality develops from the Emptiness, TLD remains connected to 4D through an **interface** of orthogonal configuration regulating two-way transfer. This transfer is asymmetric because 4D is less symmetric than TLD. Because the TLD is the informational source that generates matter in 4D, it possesses complete information about that matter. However, the 4D realm may not comprehend every piece of TLD information connected to it. All transformations, including DM, originate there, making the TLD indirectly measurable in 4D.

Emptiness to the structural asymmetry of Protospace.

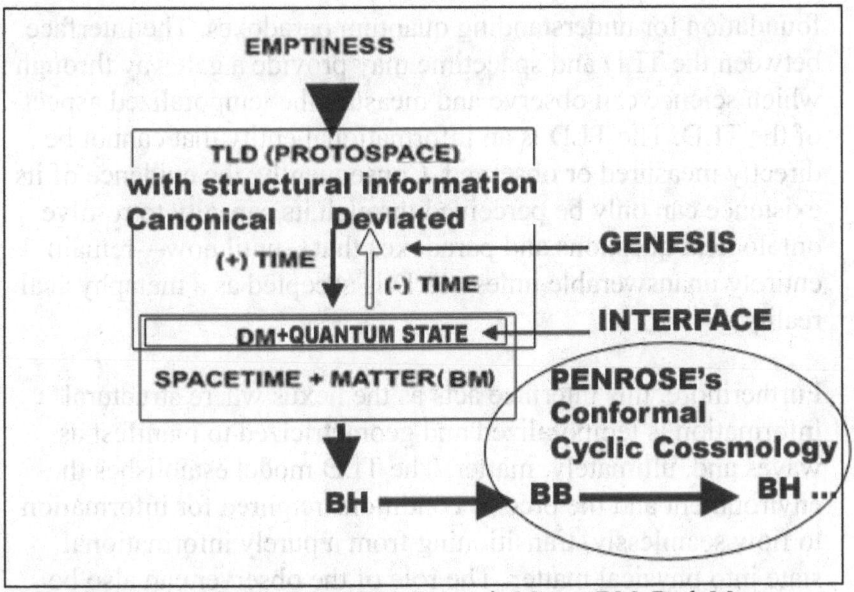

BH: Black Hole, BB: Big Bang, BM: Baryonic Matter, DM: Dark Matter

The TLD serves as the cosmic repository of *Prajñā*—the ultimate, non-local knowing. It shows that before there was a universe to be observed, the geometric architecture of observation itself already existed, encoded in the non-collinear relationship of bend and twist. Living brings are not separate from this foundation; every time a quantum wave collapses, and every time the human mind experiences a seamless, unconstrained narrative in a dream, human briefly steps outside the metric of time and touch the edges of the timeless canvas from which all phenomena flow. As a result, matter, time and information Thus, through matter, time, and information—the observable and investigable manifestations —science can gradually gain insight into the deeper layers of reality.

In summary, if Emptiness and freedom were the primordial foundations of the universe, TLD, information I, time t, and matter M emerged naturally, forming two-domain reality: physical-phenomenal spacetime and metaphysical-fundamental TLD. This model explains why spatial curvature and gravity emerged so early and proposes a deeper informational foundation for understanding quantum paradoxes. The interface between the TLD and spacetime may provide a gateway through which science can observe and measure the temporalized aspect of the TLD. The TLD is an informational entity that cannot be directly measured or observed. Consequently, the evidence of its existence can only be perceived through its capacity to resolve ontological questions and paradoxes that—until now—remain entirely unanswerable unless TLD is accepted as a metaphysical reality.

Furthermore, this interface acts as the nexus where structural information is temporalized and geometricized to manifest as waves and, ultimately, matter. The TLD model establishes the environment and the precise conditions required for information to flow seamlessly, transitioning from a purely informational state into physical matter. The role of the observer can also be naturally explained through this framework, as quantum particles are permanently registered and known, constantly observed within the foundational layer of the TLD.

$$EM \rightarrow TLD \rightarrow I_{integral} \rightarrow I_{deviated}\,(\theta,\ \kappa\,,\ \tau\,) \rightarrow t_0 \rightarrow (?)\ t_1 \rightarrow DM$$

$$t_2 \rightarrow OM$$

The concept of the Timeless Domain directly answers the foundational question: *Could a timeless domain be the foundation of all phenomena?* The answer is a resounding yes. By eliminating time t from the primordial equation and replacing it with the structural interplay of torsion τ and curvature κ, we resolve the oldest paradox in cosmology: how a dynamic universe can arise from a timeless void.
The origin of the universe was not a physical explosion of matter from a literal point in space and time. Rather, it was an **informational phase transition from a state of absolute symmetry, followed by successive cycles of black-hole formation and Big Bang expansion, with no apparent absolute beginning**

CHAPTER 1: UNRESOLVED QUESTIONS?

Summary
This opening chapter provides an overview of modern science and mathematics. Over the past two centuries, science has achieved tremendous success in describing the phenomena of the material world. However, when probing deeper into the ontological questions and the ultimate origins of these phenomena, we encounter a dense "wall of knowledge."

Consequently, many phenomena today are merely observed and successfully applied in experiments, while their underlying fundamental mechanisms remain without a unified explanation. Phenomena such as quantum entanglement, quantum superposition, the collapse of the wave function upon observation, or hypotheses regarding negative time remain open questions in modern physics.

The issue does not lie in the non-existence of these phenomena, but rather in the incompleteness of our understanding of them. In this spirit, Albert Einstein'τ famous assessment of quantum mechanics remains valid today: a theory may accurately describe a phenomenon but still fail to fully reflect the profound nature of reality.

Arising from the logical argument that the temporal domain and the timeless domain are two distinct states of existence, this book proposes that a timeless domain must exist as the foundation of the physical world. From this foundation, physical phenomena —particularly quantum phenomena—along with physical constants, mathematical structures, and even the foundational axioms of mathematics, can be understood as sequential manifestations of a deeper informational structure.

This chapter will review the fundamental open questions of physics and mathematics, thereby clarifying why it is necessary to consider a broader theoretical framework—one in which the ***Timeless Domain*** *is proposed as the common foundation for the emergence of space, time, and the material world.*

I. Why Does Modern Physics Still Face Unanswered Questions?

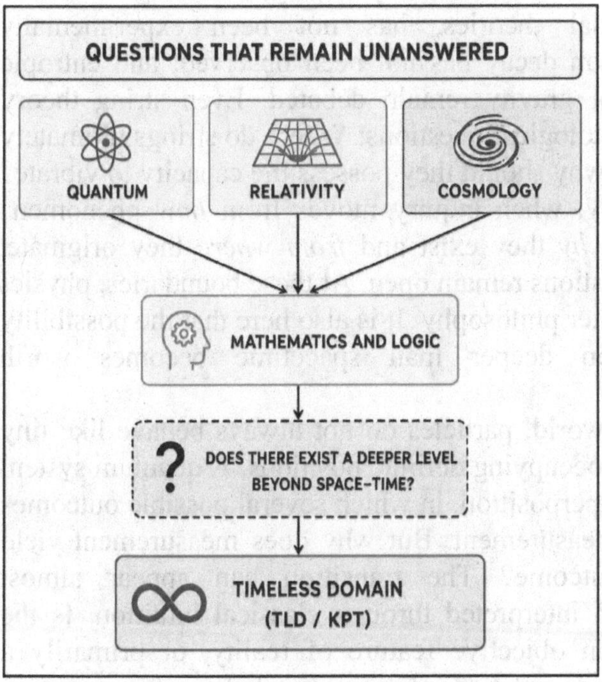

Modern physics stands among the greatest achievements of human intelligence. Quantum mechanics opened the microscopic world of atoms, elementary particles, and phenomena that transcend ordinary intuition. Einstein'τ general theory of relativity revealed that gravity is not simply an attractive force acting between objects, but can be understood as a manifestation of spacetime curvature. The Standard Model of particle physics has described with extraordinary precision the elementary particles and three—the electromagnetic, weak, and strong interactions—of the four fundamental interactions of nature, the fourth being gravity. From these theories have emerged computers, lasers, mobile phones, satellites, particle accelerators, GPS, and countless other technologies.

Yet success in description and application does not mean that humanity has reached the ultimate nature of reality. Modern physics is exceptionally powerful in describing phenomena and predicting experimental outcomes, but major questions remain unresolved. A Grand Unified Theory (GUT) has not yet been experimentally confirmed; supersymmetry, including its role in higher-dimensional theories, has not been experimentally established; proton decay has not been observed; and entropic interpretations of gravity remain debated. Even string theory leaves deeper ontological questions: Where do strings ultimately come from, and why should they possess the capacity to vibrate? More importantly, when inquiry moves from *how* phenomena behave toward *why* they exist and *from where* they originate, fundamental questions remain open. At these boundaries, physics begins to encounter philosophy. It is also here that the possibility of a foundation deeper than spacetime becomes worth considering.

In the quantum world, particles do not always behave like tiny classical objects occupying definite positions. A quantum system may exist in a superposition, in which several possible outcomes coexist before measurement. But why does measurement yield one definite outcome? The transition can appear almost mysterious when interpreted through classical intuition. Is the wave function an objective feature of reality, or primarily a mathematical representation of our knowledge of a quantum system? This remains one of the deepest interpretive questions in quantum mechanics.

Quantum entanglement is even more remarkable. Two particles that have interacted may retain correlations so profound that measurements performed on one are correlated with measurements performed on the other, even when they are separated by enormous distances. Einstein famously referred to this as "spooky action at a distance." Quantum mechanics predicts these correlations with extraordinary accuracy, yet an ontological question remains: what underlying structure allows spatially separated entities to behave as components of a single quantum system?

Quantum tunneling presents a related challenge. According to classical intuition, a particle cannot cross an energy barrier if it lacks sufficient energy. In quantum mechanics, however, there remains a finite probability of detecting the particle on the other side of the barrier. This phenomenon is routinely used in modern technology, yet philosophically it suggests that microscopic reality cannot be fully represented by ordinary spatial pictures.

The conceptual strangeness of quantum mechanics becomes particularly striking when experimental arrangements involving measurement appear to affect which properties can meaningfully be assigned to a quantum system. This gives rise to the observer and measurement problems, among the most fundamental and debated issues in quantum theory. Reports of phenomena described as "negative time" or negative interaction durations in carefully defined quantum experiments also challenge ordinary temporal intuition. Spacetime itself presents an equally profound problem. In general relativity, spacetime is dynamic: it can curve, expand, contract, and respond to matter and energy. Yet general relativity does not explain why spacetime exists in the first place; it begins with spacetime as part of its fundamental structure. Quantum mechanics, meanwhile, generally treats time as an externally supplied parameter. Thus, the two great frameworks of modern physics do not yet provide a complete account of the origin of spacetime itself.

The Big Bang brings us to a similar conceptual boundary. If the observable universe evolved from an extremely hot and dense early state, natural questions arise: What, if anything, preceded that state? In what sense could the Big Bang occur "within" anything? If physical time itself emerged with the early universe, does the word *before* retain its ordinary meaning? Such questions suggest that time may not simply be an eternally flowing background. It may instead be emergent—not an ultimate foundation, but a phenomenon arising from a deeper level of reality.

Black holes open another frontier. According to general relativity, a black hole is a region in which spacetime is curved so strongly that, beyond the event horizon, even light cannot escape to the exterior. When quantum theory is brought into the picture, however, black holes acquire temperature and entropy and are associated with Hawking radiation. This leads to the black-hole information problem: if an object falls into a black hole, is its information ultimately lost? If information is preserved, where and how is it encoded? If it is truly destroyed, fundamental principles of quantum theory appear to be threatened. A black hole is therefore not merely an astronomical object; it is also a gateway into deeper questions concerning information, spacetime, and reality.

Dark matter presents another major mystery. Astronomical observations indicate the presence of an unseen component that produces gravitational effects, influencing the motions of galaxies and the propagation of light through the universe. Yet this component does not emit or absorb light in the ordinary way and has not been directly identified as a specific particle or substance. Its gravitational influence is evident, but its fundamental nature remains unknown. This raises an important question: must gravity always be associated with ordinary matter, or could gravitational behavior reveal a structure deeper than conventional matter itself?

Dark energy is likewise poorly understood. The universe is not merely expanding; its expansion is accelerating. What drives this acceleration? Is it a cosmological constant, vacuum energy, a new dynamical field, or an indication that our understanding of gravity and spacetime remains incomplete? This question remains at the center of modern cosmology.

Even mass is not a completely closed conceptual problem. No fundamental particle with negative rest mass has been established experimentally. Phenomena sometimes described in terms of "negative mass" generally involve effective quantities arising from special dynamical or material conditions rather than a demonstrated fundamental negative rest mass. The Higgs mechanism explains how several elementary particles acquire mass through their interaction with the Higgs field. Yet deeper questions remain: Why does the Higgs field exist? Why does it possess its observed vacuum value? Why are some particles massive while the photon remains massless? The Higgs mechanism explains an essential mechanism through which mass appears, but it does not by itself answer every ontological question concerning the ultimate origin of mass.

From these questions emerges a central possibility: perhaps time should not be regarded too simply as a pre-existing background.

In ordinary experience, we imagine time flowing from past to present and into the future. Physics, however, presents several different faces of time. In general relativity, time is intertwined with space and depends on gravitational fields, motion, and cosmic geometry. In quantum mechanics, time commonly appears as an external parameter according to which quantum states evolve. In thermodynamics, the direction of time is associated with increasing entropy. Time therefore appears simultaneously fundamental and derivative: immediate and unavoidable in experience, yet possibly emergent from a deeper structure.

If time is emergent rather than fundamental, then matter, energy, mass, and even gravity might likewise not constitute the ultimate level of reality. They could represent different manifestations of an underlying informational or structural domain that does not yet belong to spacetime. From this perspective, spacetime would cease to be the absolute starting point and would instead become a stage in the realization of physical reality. What possesses time might emerge from a domain that is, at its foundation, timeless.

Mathematics itself also reveals fundamental limits. Mathematics is constructed upon axioms—statements not proved from within the system but accepted as foundations for reasoning. Euclidean geometry, for example, begins with basic postulates, from which an entire geometric world can be constructed. Yet those axioms do not prove themselves. They are starting points rather than conclusions explained by the system they generate.

The origin and ontological status of numbers present an equally profound question. Natural numbers appear self-evident: 1, 2, 3, and so forth. But what exactly is a number? Is it a construction of the human mind, or does it correspond to an objective structure of reality? Irrational numbers such as $\sqrt{2}$ cannot be represented as ratios of integers, yet arise naturally in geometry. π appears in the geometry of circles, while e emerges in growth, probability, and calculus. Mathematical constants appear almost like signatures of an underlying order. Mathematics uses them with extraordinary success, yet their ultimate ontological status remains philosophically unsettled.

Gödel'τ incompleteness theorems reveal another profound limitation. A sufficiently powerful consistent formal system capable of expressing arithmetic cannot be both complete and capable of establishing all relevant truths about itself from within its own formal resources. In particular, under the appropriate assumptions, such a system cannot prove its own consistency. There will therefore be mathematical truths that cannot be derived solely from the axioms of the system in which they are expressed. This discovery demonstrates that sufficiently rich formal systems possess intrinsic limits to self-explanation.

This observation carries important implications for the philosophy of science. If mathematics—the most precise language available to physics—rests upon foundational assumptions that are not themselves established from within every formal system, physics likewise cannot necessarily explain all of its own foundations solely through its equations. Every theory begins from assumptions, structures, laws, or parameters. When inquiry is pushed far enough, we repeatedly encounter an unexplained foundation: Why are there laws? Why numbers? Why these constants? Why spacetime? Why is there ordered structure rather than complete disorder?

The gravitational constant (G) provides a representative example. In physical equations, (G) determines the strength with which gravitational effects appear. But why does (G) possess precisely its observed value? If it were radically different, the formation and evolution of stars, planets, galaxies, and perhaps life would also be radically different. Physics can measure (G) and employ it with remarkable precision, yet current theory does not derive its observed numerical value from a universally accepted deeper principle. The same is broadly true of the speed of light (c), Planck'τ constant (\hbar), the elementary electric charge, and other fundamental constants: within our present theories, their measured values enter largely as empirical inputs.

We can therefore identify a common pattern. Science describes the phenomenal world with extraordinary success, yet deeper inquiry repeatedly encounters foundations that are not fully explained from within existing theories. Quantum mechanics confronts the measurement problem and questions of interpretation. General relativity takes the structure of spacetime as fundamental to its formulation. Cosmology requires an account of its initial or boundary conditions. Mathematics begins from axioms. Physics contains fundamental constants whose observed values are not all derived from deeper principles. Together these limitations point toward a larger philosophical question: **might there exist a foundational domain deeper than spacetime, matter, and the laws manifested within the observable universe?**

This book proposes such a hypothesis. This foundational domain will be called the **Timeless Domain (TLD)**. TLD is not proposed as another physical object located somewhere inside the universe, nor as space in the ordinary geometric sense. Rather, it is hypothesized as a **pre-spatial and timeless foundational domain** in which informational structures, relations, possibilities, and organizing principles may exist prior to their realization as spacetime, matter, energy, and consciousness.

In this framework, TLD is not presented as an experimentally established component of contemporary physics, but as a philosophical–mathematical hypothesis intended to explore what might lie beneath the assumptions upon which physical description begins. Its purpose is therefore not to replace quantum mechanics, relativity, or established physical theories, but to ask whether their unresolved foundational questions may point toward a deeper informational level of reality.

According to this hypothesis, **TLD is not intended to replace modern physics**. It does not reject quantum mechanics, general relativity, or mathematics. Rather, it attempts to place these established frameworks within a broader conceptual perspective. Modern physics describes the physical world once time has already emerged. But if time itself is an emergent phenomenon, then a deeper domain may be required to account for the emergence of time. It is at this level that **TLD—a Proto-Spatial Domain—is proposed as a deeper foundation, or possible source, of all phenomena**.

The purpose of this chapter, therefore, is to show why a hypothesis of this kind may reasonably be considered. When physics, mathematics, and cosmology encounter unresolved foundational questions, we are justified in asking a more fundamental question: **Could reality begin neither with matter nor with time, but with a timeless domain in which primordial informational structures exist prior to their realization as physical phenomena?**

If the answer is yes, then LTD would be more than a philosophical concept. It could provide a conceptual bridge between **science and metaphysics, physics and consciousness, phenomena and their underlying foundation**. From this point, we may proceed to the next chapter and ask the central question: When physics, mathematics, and cosmology all encounter unresolved foundational questions, we are justified in reconsidering a more fundamental question: **Could it be that reality begins neither with matter nor with time, but with a Timeless Domain, where primordial informational structures await realization?**

II. From Information to Primordial Knowing

When the fundamental questions of physics are pursued to their deepest level, a more profound question emerges: before matter, energy, and time came into existence, might there already have been an informational foundation? Without information, there can be no laws. Without laws, there can be no order. Without order, the universe itself could not come into being.

However, primordial information should not be understood as data stored in a computer. Rather, it is **structural information**—the fundamental relationships, possibilities, principles, and patterns that underlie all phenomena. In its timeless state, this information has not yet become matter, energy, or individual consciousness, yet it already possesses the potential to serve as the foundation of everything that follows.

The ultimate totality of this primordial information may be called **Primordial Knowing**. In the language of religion, this concept closely resembles the idea of the **Creator**. In Buddhist thought, it corresponds more closely to **Prajñā (Transcendent Wisdom)** or **Pure Awareness**. In philosophical terms, it may be understood as the intelligent foundation of reality itself.

Primordial Awarenessmay be characterized by three fundamental attributes.

The first is **Omniscience**. This is not accumulated knowledge stored in a brain, but the capacity to encompass all relationships

and all possibilities within the very foundation of reality. It is direct knowing—independent of the senses, beyond time, and unfragmented by the distinction between subject and object.
The second is **Omnipotence**. This does not signify coercive power imposed from outside, but rather the inherent capacity through which possibilities may become actualized as spacetime, energy, matter, and life.
The third is **Perfection**. It is the original goodness, order, harmony, love, and morality embedded within the foundation of reality. In this sense, morality is not merely a social convention but the manifestation of structurally coherent information within the creative foundation itself.
From this perspective, it becomes understandable why the world'τ great spiritual traditions place morality at their center. The **Ten Commandments**, the **Five Precepts**, compassion, charity, selflessness, and universal love may all be viewed as different expressions of the same underlying principle: because all existence arises from a common foundation, no being possesses the absolute right to violate the existence or dignity of another.
Throughout the remainder of this book, this foundational domain will be referred to as the **Timeless Logic Domain (TLD)**.

III. In the Book of Genesis, Light Appears on the Very First Day

In the Book of Genesis, light appears on the very first day—before the sun, the moon, and the stars are placed in the heavens. This suggests that the first "light" should not necessarily be understood as the physical light emitted by celestial bodies. Rather, it may represent a more primordial form of light: the light of order, awareness, and information.
From the perspective of the TLD, before the existence of physical matter there already existed a timeless informational foundation. When Primordial Awareness encounters pure information, the first experience is not a physical image but illumination itself. This light requires no sun, for it does not originate from material objects. Instead, it signifies the first manifestation of order emerging from the formless foundation.

Accordingly, the **light of heaven** may be understood as the experience of encountering pure information. When information remains free from distortion, Primordial Awarenessexperiences it as light, peace, and life. As information gradually differentiates into multiple levels, this primordial light is progressively manifested within spacetime as the various forms of physical light —the sun, the moon, the stars, and ultimately photons.

The same principle may also be applied to the concept of **the sound of heaven**. Primordial sound need not be understood as mechanical waves propagating through air. Rather, it may represent the experience of Primordial Awarenessperceiving the rhythm, order, and intrinsic vibration of information itself. In many spiritual traditions, this sacred sound is not heard by the physical ears but by consciousness when it reaches deeper informational levels of reality.

Thus, the light and sound of heaven are not ordinary physical phenomena. They are two complementary ways in which Primordial Awarenessexperiences pure information: **light is the manifestation of order, while sound is the rhythm of that order.**

One may therefore say that before physical light existed, there was already **the light of Awareness**; before physical sound existed, there was already **the sound of information**. The sun illuminates the visible world, whereas primordial light illuminates the very possibility of the world'τ existence.

This translation preserves the philosophical style and paragraph structure of the original while using terminology consistent with the previous chapters (e.g., **Primordial Awareness, structural information**, **Timeless Logic Domain (TLD), Pure Awareness**, and **Prajñā**). It also retains the interfaith tone without presenting these metaphysical interpretations as established scientific facts.

IV. Indications That Time Is Not the Ultimate Foundation

The idea that time is not the ultimate foundation of reality is not entirely new. Throughout the history of philosophy and physics, many thinkers have recognized that time may not exist independently, but is instead inseparably connected with motion, relationships, change, or states of awareness.

Aristotle defined time as "the measure of motion with respect to before and after." In other words, time is not a self-existing entity but rather the way in which the intellect recognizes the order of change. Without motion, and without any distinction between before and after, the very concept of time loses its foundation.

In the modern era, **Ernst Mach** argued that time should not be regarded as an absolute stage existing independently of the material world. His ideas profoundly influenced Einstein. In the theory of relativity, time is no longer a universal flow that is identical for all observers. Instead, it depends upon motion, gravity, and the observer'τ frame of reference. As summarized by the **Gravity Probe B** mission, space and time are relative and depend upon the motion of the observer.
Near a black hole, our ordinary intuition about time is challenged even more dramatically. To a distant observer, an object falling toward the event horizon appears to slow indefinitely, almost "freezing" at the boundary of the black hole. NASA describes this phenomenon as a consequence of gravitational time dilation in an extremely strong gravitational field.

Stephen Hawking further showed that near the Big Bang and in black hole physics, the classical concept of time no longer retains its familiar meaning. In his formulation of **imaginary time** and the **no-boundary proposal**, time may be treated mathematically in a manner similar to a spatial dimension, thereby blurring the conventional notion of what occurred "before the Big Bang." Pushed to the extreme conditions of a black hole, time and space appear to exchange roles. In the equation for a black hole (the Schwarzschild solution),

$$ds^2 = -\left(1 - \frac{2GM}{rc^2}\right)c^2dt^2 + \left(1 - \frac{2GM}{rc^2}\right)^{-1} dr^2 + r^2(d\theta^2 + \sin^2\theta d\phi^2)$$

the first term on the right-hand side refers to time, while the second refers to space. When r, the distance from the center of the black hole, reaches r=rs (the Schwarzschild radius of the black hole), then t=0 and space becomes infinite.

Going further inside, the causal roles of the time coordinate and the radial coordinate are exchanged. An object that is already inside the event horizon cannot "remain stationary at a fixed value of r" or return to the outside: **moving toward r=0 becomes a causally unavoidable future direction**, much as, in the outside world, we cannot stop moving toward the future.

Julian Barbour went even further by proposing that time itself is an illusion. According to Barbour, the deeper reality may consist of timeless configurations, while our perception of the passage of time arises from the "records" or "time capsules" embedded within these configurations. Although this perspective does not fully resolve the problems of consciousness and lived experience, it is significant because it pushes the question of time to its deepest conceptual foundation.

Carlo Rovelli, through the framework of **Loop Quantum Gravity** and relational physics, likewise argues that time is not a fundamental entity existing independently of physical systems. Rather, time emerges from the relationships, changes, and interactions among physical systems. In this view, there is no absolute cosmic clock—only the evolving relationships between different parts of the universe.
Within quantum gravity, the issue becomes even more profound.

The **Wheeler–DeWitt equation**, often regarded as one of the fundamental equations of quantum cosmology, contains no explicit time variable in the conventional sense. This absence gives rise to the well-known **problem of time** in quantum gravity. The **Page–Wootters mechanism** subsequently proposed that dynamics and the experience of time may emerge from quantum entanglement between subsystems within a universe that is fundamentally timeless.

Although these approaches differ in their details, they all point toward the same central insight: **time may not be the ultimate foundation of reality**. It may instead be a measure of motion, a relationship among physical systems, an effect of gravity, a structure emerging from quantum entanglement, or an experiential construct arising from memory and consciousness.

The **Timeless Logic Domain (TLD)** hypothesis continues along this line of inquiry. If time is not fundamental, then a deeper question naturally arises: **what underlies time itself?** The answer proposed in this book is that what precedes time is not "nothingness," but a timeless informational domain—the **Timeless Logic Domain (TLD)**, a foundational realm in which the primordial structures of reality exist prior to their manifestation as time, space, matter, and energy.

CHAPTER II
WHAT IS THE TIMELESS DOMAIN (TD/TLD)?

Summary

The concept of the **Timeless Domain (TLD)** may be intuitively approached through several analogies, such as the surface of a printed page, the screen of a television, or the scenes that arise within a dream. These images are not exact representations of TLD, but they help suggest that what manifests in the physical world may depend upon a deeper foundation that is not itself directly observable.

According to the hypothesis proposed in this book, TLD is a **pre-geometric domain** containing **primordial structural information** in a normative state. This information is neither matter, energy, nor a physical field. Rather, it consists of relations, organizing principles, and possibilities that exist prior to their realization as physical phenomena.

Within this model, primordial information may be extraordinarily diverse. However, because TLD is hypothesized as a domain of freedom and possibility, **deviated information** may also arise. Such deviation is proposed as the initial condition that triggers the stratification of information and, subsequently, the emergence of time, gravity, matter, and other physical phenomena.

Mathematically, informational units may be represented as **information fibers** orthogonal to four-dimensional spacetime. These fibers combine to form **fiber bundles**, and the totality of such bundles constitutes an orthogonal, timeless domain that cannot be directly observed by conventional physical methods.

At a formal level, this concept may be loosely associated with **Wick rotation**, through which the time coordinate is transformed into a coordinate with geometric properties analogous to those of spatial coordinates. In the TLD model, however, this connection is intended only as a mathematical analogy. It does not imply that TLD is identical to the Euclidean space obtained through Wick rotation.

The fiber-bundle space of TLD is hypothesized as a domain in which primordial structural information is encoded. It is **not** an extension of the eleven-dimensional spacetime of M-theory, nor does it correspond to the compactified extra dimensions of string theory. Unlike the additional dimensions considered in M-theory, which remain part of a higher-dimensional physical framework and can support physical fields and degrees of freedom, TLD is proposed as a **pre-geometric, timeless, and non-material domain** underlying the emergence of physical reality.

Within TLD:

- there is no time;
- there is no matter;
- there is no energy;
- there is no physical space.

There exist only **primordial informational relations and structures in a timeless state**, prior to their manifestation as the physical world.

I. Emptiness (EM) and the Timeless Domain (TLD/TLD)
In the model proposed here, the ultimate foundation of reality is neither matter, energy, nor mathematics, but a reality that cannot be directly described—analogous to the concept of **Emptiness (EM/Śūnyatā)** in Buddhism or the **Dao** in Daoist philosophy. EM is not nothingness.
EM is the absolute potentiality of all manifestation, while itself remaining beyond all conceptual dualities.
From EM emerges the **Timeless Domain (TLD)**, a pre-temporal domain containing the fundamental principles of consistency underlying reality.
Within the TLD, there is as yet:

- no time,
- no matter,
- no energy,
- no physical space.

Only primordial informational relations and structures exist in a timeless form.
II. TLD— A Logical Consequence If Time Is Not a Fundamental Reality
If Chapter I shows that modern physics still confronts many unresolved foundational questions, Chapter II raises a deeper question: **if spacetime is not the ultimate foundation of reality, what exists prior to spacetime?**
The hypothesis proposed in this book is the existence of the **Timeless Domain**, also referred to as the **Timeless Logic Domain (TLD)**. It is neither a space in the ordinary geometrical sense nor a domain of matter or energy. Rather, it is a **Protospace** —a foundational domain of primordial information and relations in which time has not yet emerged.
This idea begins with a simple but consequential observation: **space and time need not be fundamentally identical concepts.** Modern physics unifies them into spacetime, but this does not necessarily require them to have the same origin or to emerge simultaneously. If time is an emergent phenomenon, then one may hypothesize a domain in which time has not yet emerged. In the present model, this is the **Timeless Domain**.

A simple analogy may help illustrate TLD: consider a blank sheet of paper or a television screen before an image appears. Both the paper and the screen belong to the physical world and therefore serve only as analogies. What matters is not their physical material, but their **capacity to contain and manifest information**. An entire image can appear on a screen without the observer experiencing it as being constructed line by line. Similarly, within TLD, informational structures are conceived as complete potentialities. When the appropriate conditions arise, they become manifested in the phenomenal world.
Another analogy for understanding TLD is the world of dreams.

Time in Dreams

During sleep, particularly during **rapid eye movement (REM) sleep**, when vivid dreaming commonly occurs and skeletal muscle activity is largely suppressed, a person may see scenes, hear voices, encounter other people, and experience narratives that seem to unfold over long periods. Yet these images and sounds do not arise from the actual propagation of external light or the physical oscillation of sound waves through the surrounding space. Rather, the brain processes and presents internally generated information. To the dreamer, these phenomena appear directly, as though they were actually present.
One remarkable feature of dreams is that time often does not follow the logic of waking experience. A person may be in one city and then instantly appear in another country without experiencing any journey between them. Someone who died many years ago may appear alongside people who are still alive. Events belonging to different periods may follow one another or overlap without seeming contradictory to the dreamer. Only after awakening do we recognize that the sequence of events was not organized according to ordinary chronological order.
In this sense, dreams illustrate an intriguing possibility: **information can be experienced without necessarily being accompanied by a clear perception of physical time and space**. The dreamer does not experience light as traveling toward the eyes, does not perceive sound waves as physically propagating through space, and may not experience the passage of time as in

waking consciousness. Instead, the informational content is experienced directly.

Dreams, of course, remain products of brain activity and occur within biological and physical time. They are therefore **not TLD**. Nevertheless, they provide a useful analogy. They demonstrate that human experience does not always depend upon the conscious perception of matter continuously moving through physical space. Similarly, in the TLD model, primordial information does not exist as light, sound, or matter, but as complete informational structures. Only when these structures are realized within spacetime do they manifest as the physical phenomena familiar to us.

A particularly subtle feature of dreaming is that the sense of time is not always the same. In many dreams, the dreamer has little awareness of temporal flow. Scenes may change abruptly and events may lack a clear chronological sequence, making the experience of time seem almost absent. In other cases, however, the dreamer may still experience time as passing—quickly or slowly—and retain some sense of before and after. Within the terminology proposed in this book, this may be regarded as a form of **"thin time,"** in which temporal succession has begun to appear but has not acquired the completeness characteristic of waking consciousness.

Altered temporal experience may also occur during transitions between sleep and wakefulness and can become particularly noticeable in **lucid dreams**. In a lucid dream, the dreamer becomes aware that they are dreaming and may sometimes deliberately think, speak, or influence events within the dream. The experience of temporal continuity may consequently become clearer and more explicit.

From the perspective proposed in this book, these states may serve as analogies for different degrees of temporal manifestation: from an apparently weak or nearly absent sense of time in many dreams, through **thin time** in transitional dream states, to a more explicit experience of temporal succession in some lucid dreams.

Neuroscience supports the broader observation that **subjective time perception can vary substantially across sleep and dream states**. However, interpreting these variations as different degrees of the *presence of time itself* is a philosophical and theoretical interpretation proposed in this book, rather than an established conclusion of neuroscience. The analogy is intended to illustrate the hypothesis that the manifestation of time may be **graded or continuous**, rather than restricted to two absolute states—time either being present or absent.

According to this view, the 4D world resembles a screen upon which the information of TLD becomes manifested as images, sounds, matter, and motion, whereas TLD represents the timeless foundation in which information exists prior to assuming the forms of the phenomenal world.

Because there is no time within TLD, there can be no physical distinction between before and after. There is no past, present, or future in the ordinary temporal sense. Information is therefore conceived as immediately present—not because it travels at infinite velocity, but because **there is no temporal interval by which such motion could be measured**.

In this sense, the timeless presence of information in TLD may be compared, at the level of philosophical analogy, with notions such as the **"unborn," "unconditioned," "eternal," or "deathless"** found in several Eastern philosophical traditions: neither arising nor ceasing and not governed by temporal succession.

If there is no time, there can likewise be no motion in the physical sense. Without motion, concepts such as **speed, velocity, acceleration, and trajectory** cease to have their ordinary meaning. And if there is no change of physical position, then space itself cannot yet exist in its familiar physical sense. TLD should therefore not be understood as a three-dimensional "space," but rather as a **Protospace**—a foundation that has not yet differentiated into coordinates, distances, directions, and boundaries.

From the standpoint of physics, energy, mass, and temporal evolution are defined within a physical theoretical framework. In Einstein'τ celebrated relation,

E=mc2,

mass and energy are related once physical quantities and spacetime descriptions are already meaningful. In the TLD hypothesis, before physical time and spacetime have emerged,

there is therefore no **physical energy**, no mass, and no elementary particles in their conventional physical sense. TLD is not an "invisible universe" containing subtler forms of ordinary matter; rather, it is hypothesized as a **pre-material domain**.
A similar distinction applies to consciousness. Knowledge and cognition in their ordinary biological sense depend upon neural activity, memory, and the functioning of the brain. Before biological brains exist, there can be no individual biological consciousness, no embodied cognitive subject, and no living organism in the conventional sense. TLD should therefore not be conceived as a place inhabited by individual beings, but as a foundational domain **prior to the emergence of individuality itself**.

What Does the LTD Contain?

According to the model proposed in this book, the TLD contains only **primordial structural information**. Here, *information* does not refer to data stored in a computer or encoded in physical media. Rather, it refers to fundamental relations, organizing principles, and latent possibilities that exist prior to their manifestation as physical phenomena. Within the TLD, however, these complementary distinctions do not yet exclude or oppose one another, because they have not yet entered the dualistic framework of space, time, and physical manifestation.
From this perspective, apparent opposites are not originally contradictory. They are simply different potential expressions contained within a deeper, unified informational order. Only when structural information is temporalized and manifested through the interface into four-dimensional spacetime do these latent distinctions become the dualities that characterize the physical world.
This information may be extraordinarily diverse and nearly infinite in its properties, pre-forms, quantities, as well as in its structural methods and mechanisms. It may be regarded as the informational precursor of all structures and living beings that later emerge. For example, the structural principles that may manifest as DNA in the physical world already exist here in the form of informational pre-forms.

Likewise, the structural information underlying every living being is already present within TLD/TLD, **simultaneously rather than sequentially**. This state may be compared to grains of sand of different sizes in a desert: although they differ from one another, they all coexist on the same level, with none intrinsically coming before or after another in temporal order.

It contains the primordial structural information from which all later distinctions may arise. These include complementary possibilities such as strong and weak, positive and negative, yin and yang, addition and subtraction. Yet these opposites do not exclude one another because they have not yet entered the dualistic world of spacetime. At this stage they exist only as structural possibilities. In this primordial condition there is neither hatred nor goodness and evil, neither sin nor conflict, because moral categories arise only after time, action, and choice become possible. For this reason, the TLD is regarded as the domain of **structurally coherent information**—information that has not yet deviated from its original state of equilibrium.

These informational units may combine into structures of increasing richness without being constrained by" size" or "distance". The totality of these informational structures may be envisioned as **Prajñā (Transcendent Wisdom)**, **Omniscience**, or **Primordial Knowing**—not the knowledge possessed by an individual mind, but the intrinsic capacity for knowing embedded within the very foundation of reality.

Where does the TLD exist?

This is a natural question, but it is also the one most likely to lead to misunderstanding. If the TLD is regarded as a location somewhere within the universe, then it has already been placed inside spacetime. In reality, the TLD does not exist "somewhere" in the geographical sense. Rather, it is the foundation that makes the very notion of "where" possible.

One analogy may help illustrate this idea. Imagine a fish living at the bottom of a lake. The fish sees only the shadow of a bird flying overhead, projected through the water onto the lakebed. For the fish, that shadow is the directly observable phenomenon. It has no

awareness that above the water lies a much larger three-dimensional world.
If the bird flies exactly perpendicular to the water'τ surface, its shadow appears as an almost stationary point. From the fish's perspective, the temporal aspect of the bird's motion nearly disappears. Only when the bird flies at an oblique angle—and sufficiently close to the water'τ surface for its shadow to become visible, while the bird itself remains unseen because of the refraction geometry—does the shadow move across the lakebed, allowing the fish to perceive the passage of time. To perceive the bird itself rather than merely its shadow, the fish would have to change its frame of reference by a ninety-degree rotation. In mathematical language, this may be viewed as a rotation into a higher orthogonal space.
In this analogy, the bird'τ world is not itself the TLD, but it symbolizes a deeper level of reality. To the fish, only the bird'τ shadow is accessible. Likewise, the four-dimensional world in which we live may itself be only the "shadow" or projection of deeper informational structures existing within the TLD.

According to the TLD model, information within the Timeless Domain may be conceived as existing in two conceptual states.
The first is **structurally coherent information**. This is a state of perfect equilibrium, free from delay and entirely independent of time. Information is present immediately and does not generate physical phenomena.
The second state is **deviated information**. This deviation does not imply that matter or energy has already emerged. Rather, it represents only a very subtle difference within the informational structure, expressed not as a **faintness or partial obscuration in its capacity for manifestation**. It is precisely this faintness that constitutes the **"angle" of deviation**, opening the possibility for the emergence of time.
In other words, the spontaneous appearance of deviated information differentiates the Timeless Domain into two conceptual regions: one consisting of structurally coherent information, and the other consisting of information possessing

the potential to generate time. Within this second region, time does not yet exist in actuality but is already present as a latent possibility. When this potential becomes actualized, physical time emerges. With time come space, energy, matter, motion, biological consciousness, and ultimately the entire four-dimensional universe that we observe.

From this perspective, the material universe is not the starting point of reality but the final outcome of a progressive realization of information. Spacetime is therefore not the ultimate foundation of reality but merely the level at which a deeper domain becomes manifest. Accordingly, the Timeless Domain is not proposed as a replacement for modern physics. Rather, it is offered as a philosophical and conceptual framework for explaining how time, matter, and the laws of physics themselves may emerge from a fundamentally timeless reality.

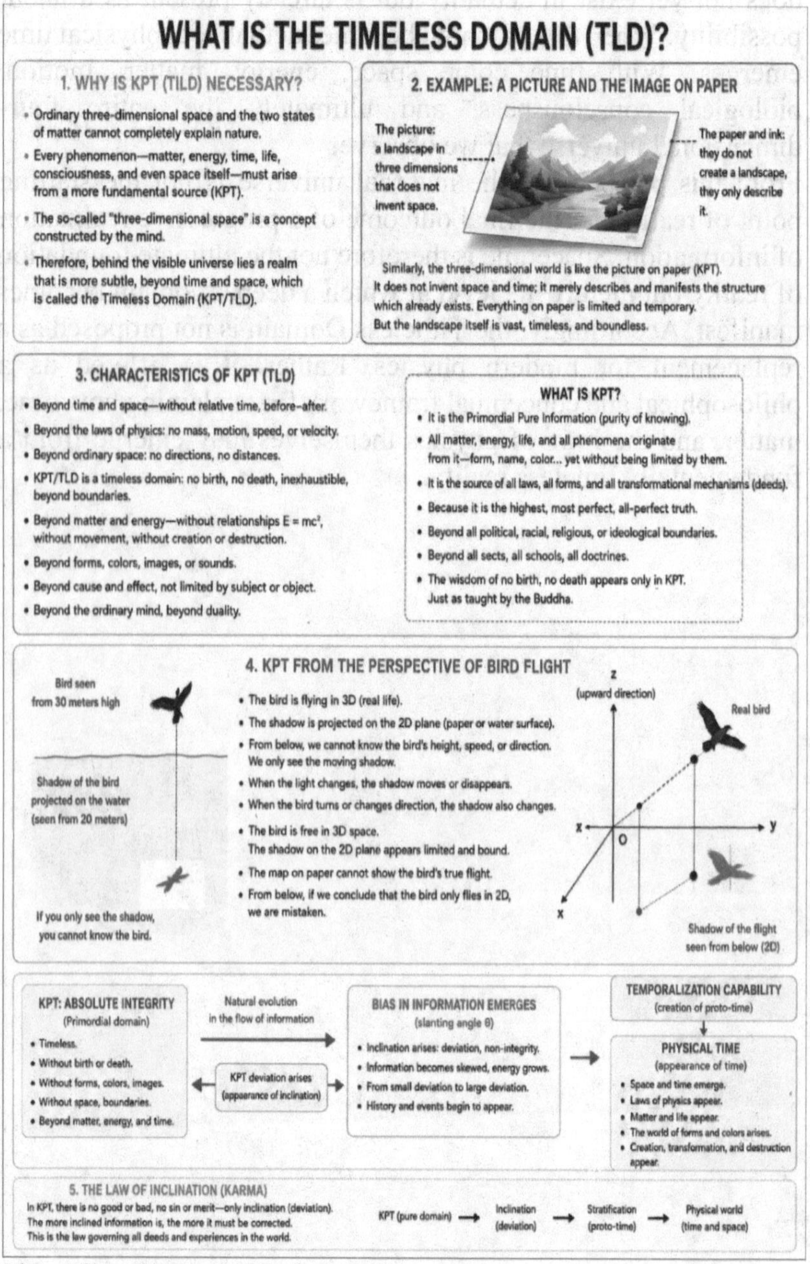

Why Can Deviated Information Arise?

In this framework, spacetime is regarded as a domain in which information becomes physically manifested. Spacetime is therefore not considered the ultimate foundation of reality, but rather a manifest layer arising from a deeper domain. LTD is consequently not proposed as a replacement for modern physics, but as a philosophical and conceptual foundation for exploring how time, matter, and physical laws might emerge from a timeless reality.

According to the TLD hypothesis, the possibility of **deviated information** arises from the existence of **freedom** as a primordial property of reality. Just as integrity, harmony, or an orientation toward the good may be regarded within this model as fundamental characteristics of TLD, freedom may likewise constitute one of its primordial attributes. Because freedom exists, informational structures possess not only the possibility of developing in accordance with their normative state, but also the possibility of deviating from it.

Without freedom, every process would be completely deterministic. There would then be no genuine choice and, within the logic of the model, no meaningful possibility of error, accident, or informational deviation. Conversely, once freedom is postulated as a primordial property of reality, the possibility of deviated information follows naturally. In other words, deviated information is not proposed as a separately created entity, but as a possible consequence of choice within a system that permits freedom.

III. TLD as a Fiber-Bundle Space, Not a Supersymmetric Space of M-Theory

Historical Background: Fiber-Bundle Space—from Mathematical Structure to a Foundation of Modern Physics

The concept of a **fiber bundle** did not emerge suddenly from a single work. It developed through a long intellectual process that began with differential geometry, matured within topology, and eventually became one of the fundamental mathematical languages of modern physics.

The deeper origins of this idea can be traced to the work of Gauss and Riemann on curved spaces. Gauss demonstrated that a surface can possess intrinsic curvature without requiring its geometry to be defined solely through an embedding in a higher-dimensional space. Riemann extended this insight into the concept of a manifold: a space may locally resemble Euclidean space while possessing a much more complex global structure. Geometry thereby moved beyond the study of objects situated *within* space toward the study of the structure of space itself.

Over a manifold, one may associate with each point an additional space—for example, a vector space, a symmetry group, or a space of internal states. The collection of these associated spaces can form a **fiber bundle**. Mathematically, a fiber bundle may be represented by a projection $\pi \rightarrow M$

with the fiber over a point (x in M) given by $\boxed{F_x=\pi^{-1}(x)} \cong \boxed{F}$

where (M) is the **base space**, (E) the **total space**, and each point (x\in M) is associated with a fiber (F_x). Locally, the total space resembles a product

$\boxed{F_x=\pi^{-1}(x)} \cong U\,.\boxed{F}$

although globally the bundle may possess nontrivial twisting or topology that prevents it from being represented as the simple product (M. F).

Fiber-bundle theory took its mature form during the twentieth century through the work of **Hassler Whitney, Charles Ehresmann, Norman Steenrod**, and many other mathematicians. Whitney developed vector bundles and characteristic classes; Ehresmann contributed fundamentally to the geometric formulation of connections; and Steenrod systematized the subject in his classic work *The Topology of Fibre Bundles*. Fiber bundles consequently became a powerful bridge connecting differential geometry, topology, algebra, and analysis.

The philosophical significance of the fiber-bundle concept lies in the way it changes our intuitive understanding of space. Space need no longer be conceived merely as an empty, homogeneous, and passive stage. Each point of a base space may be associated with its own internal mathematical structure. Reality may therefore be characterized not only by the location of a point, but also by the possible directions, states, transformations, and symmetries associated with it.

In differential geometry, for example, each point (x) of a smooth manifold (M) is associated with a **tangent space** (T_xM), containing the possible tangent directions at that point. The collection of all tangent spaces forms the **tangent bundle**

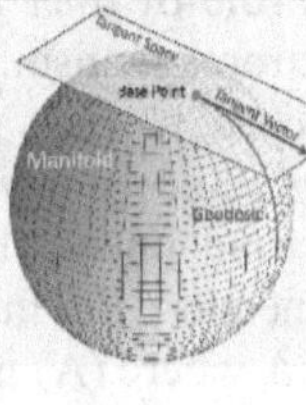

$$TM = ¿ x \in M \{x\} \times T_x M$$

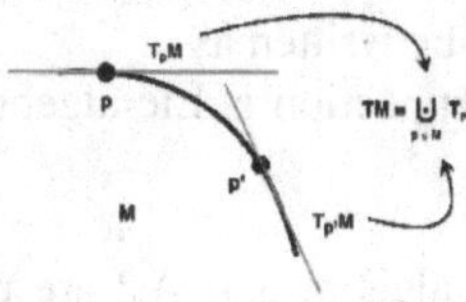

This structure provides the mathe*Type equation here.*matical framework for defining tangent vectors, vector fields, derivatives, trajectories, and related geometric quantities. More generally, the concept of a **connection** allows structures at neighboring points to be compared and transported, while curvature measures the failure of such transport to behave as it would in a flat geometry.

In physics, an important conceptual transition occurred when it became clear that physical fields could be formulated not merely as quantities distributed throughout spacetime, but as geometric structures associated with bundles over spacetime. This viewpoint became especially powerful in gauge theories.

In general relativity, spacetime is modeled as a four-dimensional differentiable manifold equipped with a Lorentzian metric. Each point possesses a tangent space, while the metric determines spacetime intervals and the Levi-Civita connection provides a rule for parallel transport between neighboring tangent spaces. The curvature associated with this connection describes the geometry of gravitation.

Thus, the mathematical language of fiber bundles provides a natural framework in which a physical point may possess not only a position in spacetime but also additional geometric or internal structure. This idea will later be adapted in the TLD model to represent **information fibers and their organization into informational fiber bundles**. Such an adaptation, however, is a hypothesis of the TLD framework and should be distinguished from the established use of fiber bundles in differential geometry and modern gauge theories.

Further Reading

The role of fiber bundles becomes even more explicit in **gauge theories**. In electromagnetism, an internal phase associated with the group U(1) may be attached to each point of spacetime. The electromagnetic potential can be interpreted geometrically as a **connection** on a principal bundle, while the electromagnetic field strength is represented by the curvature of that connection. For the Abelian (U(1) theory, this relation takes the form

$F=dA$ where (A) is the gauge potential, or connection one-form, and (F) is the field-strength two-form. Locally, the connection may be written as

$A=A_\mu dx = A_\mu{}^a$ In a general Yang–Mills theory, the connection is Lie-algebra-valued: $A_\mu=A_\mu{}^a T_a dx^\mu$

or equivalently, $A_\mu=A_\mu{}^a T_a$

where (T_a) are generators of the corresponding Lie algebra and $A_\mu{}^a$ are the gauge-field components.

In **Yang–Mills theories**, this structure is extended to non-Abelian gauge groups such as (SU(2)) and (SU(3)). The curvature then takes the form $F=dA+A \wedge A$

Here, (\wedge) denotes the antisymmetric **wedge product of differential forms**, while the non-Abelian character arises because the Lie-algebra generators generally do not commute:

Consequently, the term (A\wedge A) does not generally vanish, and the gauge fields can interact with themselves. This nonlinearity is a fundamental feature of the gauge theories underlying the weak and strong interactions.

The Standard Model of particle physics is based on the gauge symmetry $[SU(3)_C \times SU(2)_L \times U(1)_Y]$

and may be formulated geometrically in terms of connections associated with the corresponding gauge structures. The gluons and the electroweak gauge bosons arise as quantum excitations of these gauge fields; after electroweak symmetry breaking, the familiar photon and ($W^- W^+$ and (Z^0) bosons emerge from the electroweak sector.

A profound implication of this formulation is that modern physics has partially blurred the conceptual boundary between **force and geometry**. A gauge field need not be viewed merely as an additional entity placed into spacetime from the outside. Geometrically, it can be represented through the way internal fibers at neighboring spacetime points are related by a connection. In this sense, physical interactions can be understood, at least at the mathematical level, through the **connection and curvature of fiber-bundle structures**.

Fiber bundles also play an important role in **string theory and higher-dimensional models**. Extra spatial dimensions may be compactified on geometrically complex spaces, including Calabi–Yau manifolds in important classes of string compactifications. Gauge and matter structures can then be described using vector bundles or related geometric constructions over these compact spaces. The geometry and topology of these spaces and bundles can influence the particle spectrum, charges, symmetries, and interactions that appear in the effective four-dimensional theory.

From a philosophical perspective, these developments suggest a significant transformation in our conception of physical reality. In classical mechanics, space serves primarily as the arena in which objects exist and move. In general relativity, spacetime itself becomes dynamical and curved. In gauge theory, each point of spacetime may additionally be associated with an **internal space** carrying symmetry information. Physical reality is therefore increasingly described not merely in terms of separate objects, but through **relations, symmetries, connections, and curvature.**

Nevertheless, in conventional mathematics and established physical theories, fiber bundles are generally constructed **over a base manifold**. In physical applications, spacetime—or a manifold related to spacetime—often serves as the base, while fibers represent internal mathematical degrees of freedom associated with its points.

This distinction is important for the TLDhypothesis developed in this book. Conventional gauge theory uses fiber-bundle geometry to describe structures **over physical spacetime**. TLD, by contrast, proposes informational fiber structures at a deeper, pre-geometric and timeless level from which spacetime itself is hypothesized to emerge. The former belongs to established mathematical physics; the latter is the theoretical extension proposed in this work.

TLD as a Foundational Informational Fiber Space

The TLD model proposes a deeper conceptual reversal. In conventional gauge theories, fiber structures are formulated over a spacetime manifold. In the present hypothesis, however, the informational fiber structure does not arise *over* spacetime. Rather, it is proposed as a deeper structural domain existing **prior to spacetime**, from which a four-dimensional physical reality may subsequently emerge. Four-dimensional spacetime, matter, and physical fields would then represent temporalized and geometrized manifestations of a more fundamental informational structure.

Under this interpretation, a fiber no longer represents only a vector space, symmetry group, or other conventional mathematical fiber. Within the TLD model, it may represent an elementary relation, a structural possibility, or a component of primordial information. The **tilt, curvature, and twist** of such fibers may then serve as modeling variables describing how information departs from its state of integrity and progressively gives rise to connections, curvature, fields, and ultimately matter.

Fiber Space as a Domain of Potential Dimensions

If the foundational TLDis timeless and intrinsically static, why does the manifested world appear dynamic and extraordinarily diverse?

Within the proposed model, this possibility is associated with the **fiber structure** of the underlying informational domain. Informational fibers may be represented as orthogonal to the corresponding structures manifested in four-dimensional spacetime. These fibers may encode potential intrinsic properties that, after physical manifestation, appear as quantities such as charge, mass, quantum phase, or probability structures.

The total organization of these fibers constitutes what will here be called **Informational Fiber Space**. At the TLD level, this structure is not imagined as physically moving through time. Rather, it is conceived as a complete, timeless structural configuration.

From this perspective, the apparent **flow of time** may be modeled as the ordered connection of a continuous sequence of **sections** through informational fiber space. What is experienced within 4D as change and temporal succession may therefore correspond, at the deeper level, to an ordered realization of different sections of an underlying timeless structure.

Accordingly, the TLD model does not identify the fundamental domain of reality with the higher-dimensional spacetime considered in M-theory. Instead, it proposes an **informational fiber space**. This is not a physical geometric space in the conventional sense, but a **pre-geometric domain** in which only primordial informational relations and structural possibilities exist.

The fundamental characteristics assigned to this domain are **integrity** and **orthogonality**. Each informational structure exists according to its intrinsic nature, prior to the constraints of physical time, space, energy, or motion. This integrity constitutes the primordial order, coherence, and internal consistency of the TLD domain.

To distinguish this foundational domain from ordinary four-dimensional spacetime, and particularly to emphasize its proposed orthogonality to 4D, it may therefore be represented conceptually as an **informational fiber space**. Its "fibers" are not physical strings or material filaments; they represent informational relations organized within an internally coherent structure.

In the TLD hypothesis, this informational fiber space is **orthogonal to four-dimensional spacetime**. This proposed orthogonality is used to explain why TLD cannot be directly detected through ordinary physical measurements or sensory observation. The material world accessible to observation would represent only the manifested form of these deeper informational structures after they pass through an **interface domain**, where temporal ordering begins to emerge and physical processes become activated.

Distinction from M-Theory

This conception differs fundamentally from M-theory. M-theory is formulated within a higher-dimensional physical framework, conventionally involving eleven-dimensional spacetime. Its additional dimensions remain geometric degrees of freedom within the theory, and different compactifications can produce lower-dimensional effective physics. In related string-theory constructions, compact internal spaces—including Calabi–Yau manifolds in important classes of models—can influence the resulting fields, symmetries, particles, and interactions.

The inability to observe such additional dimensions directly is therefore conceptually different from the proposed invisibility of TLD. In higher-dimensional physical theories, additional dimensions may be compactified or otherwise inaccessible at ordinary energy and length scales. TLD, by contrast, is hypothesized to be inaccessible because it is **not itself an additional physical dimension of spacetime**.

In the TLD model, **informational fiber space is neither a compactified geometric space nor an extension of physical spacetime**. It is proposed as a pre-geometric domain logically prior to geometry and to the emergence of physical time. From this primordial informational domain, geometry, spacetime, matter, and physical fields are hypothesized to become progressively manifested through the interface domain. For this reason, TLD may be described within the model as a **foundational domain** rather than an additional physical space.

In concise terms:

M-theory extends the geometry of the physical world, whereas TLD proposes an informational foundation prior to physical geometry.

According to this hypothesis, geometry is not the ultimate level of reality. It is an emergent manifestation of more fundamental informational relations within TLD. The proposed hierarchy may therefore be expressed schematically as

Primordial Information→Geometrization+ Tempȯalization→Spacetime→Fields and Matter

From this perspective, TLD is not intended to replace M-theory. Rather, it is proposed at a conceptually deeper foundational level:

information precedes geometry, and geometry precedes the physical manifestation of matter.

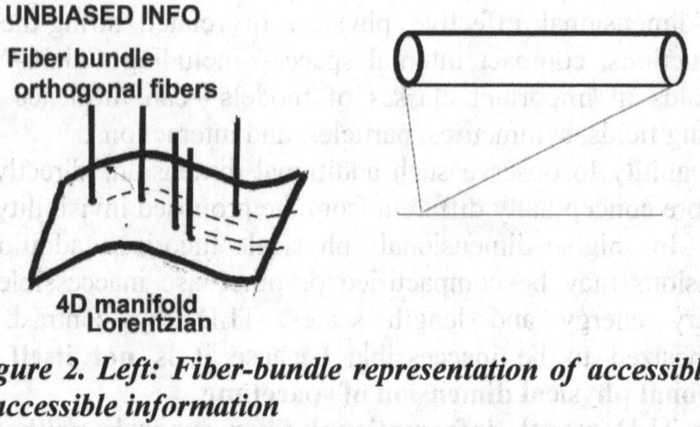

Figure 2. Left: Fiber-bundle representation of accessible and inaccessible information

*A **fiber-bundle** diagram illustrating the relationship between **physically accessible** and **physically inaccessible information****.
The four-dimensional base manifold represents* ***spacetime****, in which each point corresponds to information that can be accessed

through interaction and measurement. At each point of the base manifold, an associated* ***fiber*** *represents informational degrees of freedom that cannot be directly accessed within 4D spacetime.*

An informational fiber orthogonal to the base represents ***unbiased information*****, remaining separated from temporal dynamics and making no contribution to sequential manifestation.*

Right: *The eleven-dimensional spatial structure: the extra dimensions are compactified within a tube whose radius is on the order of the Planck length.*

.

Diversity Within Structural Integrity

In the primordial Timeless Domain (TLD), individuality does not yet exist in the sense found within the phenomenal world. All information remains in a state of **structural integrity**—timeless, undifferentiated, and not yet stratified into varying degrees of deviation. Because there is no time, there is no history, no personal memory, and no distinction between subject and object. Consequently, there is not yet an individual self or ego in the ordinary sense.

This, however, does **not** imply that all information within the primordial TLD is completely homogeneous or devoid of distinction. According to the TLD model, even within a state of perfect structural integrity there may exist different informational structures. These differences, however, are entirely harmonious, mutually complementary, and free from conflict or mutual exclusion. They represent **diversity within unity**, rather than separation through opposition.

From this perspective, the primordial TLD may already contain distinct **archetypes** or **primordial proto-individuals**, each possessing its own intrinsic characteristics. Yet all of these exist in a state of perfect structural integrity. Difference is therefore no longer the source of division but rather the expression of richness within a single unified foundation.

This perspective also helps explain why many of the world'τ major religious traditions describe diversity even within realms regarded as perfectly pure. In Christianity, for example, angels and the celestial hierarchies fulfill different roles while remaining united in the divine order. In Buddhism, the Buddhas and Bodhisattvas are all fully enlightened, yet each manifests different vows, skillful means, and compassionate activities. These distinctions do not signify superiority or opposition; rather, they express the richness of one and the same ultimate truth.

According to the TLD model, it is only when information begins to deviate from its state of structural integrity that this harmonious diversity becomes stratified. At that point, distinctions of separation, degree, history, karma, and eventually individual beings within spacetime begin to emerge. It is therefore essential to distinguish between two fundamentally different concepts:

- **Diversity within structural integrity** is an intrinsic characteristic of the primordial TLD.
- **Individualization through stratification** is a consequence of deviated information.

Stratification does not create the intrinsic value of individuals. Rather, it creates the conditions within which learning, moral development, and self-correction become possible. When information returns to its state of structural integrity, separation disappears, yet the richness of informational structures need not vanish. What is restored is **perfect harmony**, not absolute uniformity.

In this framework, **structural integrity** signifies **perfect harmony despite diversity**. Because no matter, images, or physical forms yet exist within the primordial TLD, structural integrity is not a physical property but an informational and ethical one. It denotes complete coherence, mutual compatibility, and perfect accordance with the moral order embedded within the foundation of reality.

Informational Fiber Space — The Foundation of Spacetime Phenomena and Their Projection into 4D Spacetime

From the perspective of the TLD model, **informational fiber space** may be regarded as the pre-geometric foundation underlying phenomena manifested within spacetime. In this framework, each informational structure, represented as a fiber, may produce two fundamentally different forms of projection when mapped into the four-dimensional physical world.

(a) Information in a State of Integrity

Informational structures that remain in their state of integrity are assumed to remain orthogonal to four-dimensional spacetime. When projected onto 4D, their projection may reduce to a point or to a structure with no physical degree of manifestation. Because they possess no tilt, curvature, or twist, they generate no observable physical phenomenon and may therefore be regarded as **unmanifested within the 4D world**.

Symbolically, this idealized state may be represented as

$$\theta = 0, \tau = 0, \kappa = 0$$

corresponding to a state of informational integrity with no physical manifestation.

(b) Tilted, Curved, and Twisted Information

When an informational structure deviates from its state of integrity, by contrast, its projection into four-dimensional spacetime begins to acquire physically manifest informational content. In the proposed model, three geometric variables—**tilt (θ), curvature (κ), and twist (τ)**—are used to characterize how such informational deviation may eventually become manifested as matter, fields, and individualized structures within the phenomenal world.

The proposed transition from TLD into spacetime is therefore not merely a geometric projection. It is also hypothesized as a process through which **time is introduced into informational structure**. This process may be called **temporalization**. Through temporalization, deviations that remain latent within the timeless domain become manifest as distinguishable structures, events, and individuals in the physical world.

According to the TLD hypothesis, the purpose of this process is not simply to generate a separate physical universe. Rather, manifestation provides a domain in which informational deviations can become identifiable, experienced, and progressively transformed.

Within the broader philosophical framework of the model, geometric deviations of information may also be interpreted as expressions of departure from an **original normative or ethical**

state. TLD itself is hypothesized to contain neither matter, energy, nor physical form in the conventional sense, but primordial informational structures characterized by integrity, order, coherence, and truthfulness. Informational deviation may therefore be interpreted at two related levels: geometrically, as **tilt, curvature, and twist**; and philosophically, as a departure from primordial ethical and spiritual integrity.

This interpretation has an interesting conceptual resonance with Buddhist philosophy. In Buddhist thought, phenomenal reality is understood as dependently arisen and therefore not as an independently existing ultimate reality. Some Buddhist traditions also speak of **Suchness (Tathatā)** as a way of referring to reality beyond conceptual fabrication. From the perspective proposed here, spacetime reality may analogously be understood as a manifested representation of deeper informational structures originating in a timeless domain.

The comparison, however, should remain an **analogy rather than an identification**. The TLD model is a speculative informational and geometric framework, whereas Buddhist concepts such as dependent origination and Suchness belong to a distinct philosophical and contemplative tradition. The purpose of the comparison is therefore not to equate the two systems, but to explore a possible conceptual bridge between **informational geometry, the emergence of physical phenomena, and Buddhist philosophy**.

THE INTERFACE DOMAIN

At this point in the TLD model, a certain similarity can be recognized between the **fiber-bundle space of TLD** and the **extra spatial dimensions of eleven-dimensional M-theory**. Neither can be directly perceived by the senses or through ordinary measurements within the 4D world. However, this similarity remains only at the **formal and phenomenological level**, while also suggesting that the two structures may share certain limited physical or mathematical characteristics.

In the TLDmodel, **fiber-bundle space** is hypothesized as a pre-geometric informational structure that does not necessarily possess the nature of physical space in the ordinary sense. By contrast, in M-theory, the extra dimensions are components of the higher-dimensional spacetime structure required by the theory. Therefore, the similarity between the two models should be regarded as a **suggestion of the possible existence of degrees of freedom beyond the direct observational reach of the 4D world, as well as of a transitional mechanism through which information may be filtered and transferred between the two domains.** It should **not be regarded as evidence that TLD is identical to the extra dimensions of M-theory**.

Chapter III

CHAPTER III: HOW DOES TIME EMERGE? THE THEORY OF INFORMATION PRE-GEOMETRIZATION AND TEMPORALIZATION INTO MATTER (IPTM)

Summary

According to the hypothesis proposed in this book, the ***Timeless Domain (TLD)*** *is a* ***pre-geometric domain*** *containing only* ***primordial structural information****, prior to matter, energy, and time in their physical sense. Mathematically, such information may be represented by abstract geometric structures, such as* ***information fibers*** *and* ***fiber bundles****, forming a pre-geometric informational space proposed as the foundation underlying physical phenomena.*

In its state of integrity, information exists as a coherent whole, not yet divided into the oppositional pairs characteristic of the dualistic phenomenal world. Symmetrical structures do not annihilate one another but coexist within a harmonious order. Within the philosophical interpretation of this model, such a state corresponds to the foundation of ***truth, goodness, and beauty****, or, more broadly, to universal ethical principles that have been expressed in different philosophical and religious traditions through concepts such as* ***Suchness, the Tao, God, or the Creator****. These concepts are invoked here as philosophical analogies rather than as equivalent doctrines.*

TLD, however, is hypothesized as a domain of ***freedom and possibility****. Consequently, informational deviation or asymmetry may arise. A difference then appears between the existing informational state and its state of integrity, creating within the model a requirement for stratification, learning, and self-correction. The IPTM hypothesis proposes that this requirement is the origin of* ***time****.*

Time, in this interpretation, emerges as part of the process through which informational order is progressively re-established after primordial symmetry has been disturbed. The need for correction activates a latent ordering capacity already inherent within TLD.

Time is therefore not treated as an independently existing substance. Rather, it represents a marker of ***mechanism for ordering, organizing, and stratifying information****, providing the conditions under which deviated information may progressively transform toward a state of integrity. This process becomes active when informational deviation crosses a specified threshold, represented in the model by an index measuring departure from the normative state.*

The IPTM model proposes that the ***pre-geometrization of information and its progressive temporalization into matter*** *proceeds through three stages of time:*

(t_0): Proto-Time

At the stage of ***(t_0)****, or proto-time, informational deviation remains small and within an acceptable range relative to the normative state. Time at this level exists only as a capacity for* ***ordering, classification, and informational registration****. There is not yet physical energy, physical fields, or matter.*

Thus, t_0: Ordering f nformation without p hysical atter . ‖ 1. t_1 *m*

inimal Time

When informational deviation exceeds a critical threshold, the requirement for correction becomes more pronounced, with its intensity increasing as the degree of deviation increases. This initiates t_1*, or minimal time.*

This stage corresponds in the IPTM model to the emergence of the ***Interface Domain****, where informational structures begin to manifest curvature and pre-material effects.* ***Dark matter is hypothesized in this model to belong to this transitional stage****, rather than to the fully materialized* t_2 *domain.*

Schematically,

[¿t_1: ‖ 1"{*Information*} ‖ 1 ‖ 1"{*Curvature*} ‖ 1 ‖ 1"{*Pre-mater*

(t_2): Physical Time

At the stage of t_2*), time becomes the fully developed physical time associated with four-dimensional spacetime. At this level, quantum fields, elementary particles, and the observable material world emerge.*

The proposed sequence can therefore be summarized as

TLD t_0 t_1 t_2 4 *D Physical Reality}*

where t_0) represents informational ordering, (t_1) the interface and pre-material stage, and t_2 physical time and material manifestation.

The Three Structural Variables of Materialization

Within IPTM, deviated information is represented by three fundamental variables, which also function as three proposed structural codes underlying material manifestation:

- ***θ** — the degree of **tilt or deviation** from the state of informational integrity;*
- ***(κ)** — the **curvature of information**, hypothesized to be related to the emergence of gravity and mass;*
- ***(τ)** — the **twist of information**, hypothesized to be related to particle spin and to the degree of manifestation of material structure.*
- ***ρ-** density of information, ℓ "length of infornation related to the size*

The state of informational integrity may therefore be represented ideally as $\theta=0, \kappa=0, \tau=0$

while departures from this state generate progressively more complex informational geometry.

*Within this framework, the manifestation of matter is not regarded as an end in itself. Rather, it is interpreted as a consequence of the requirement for information to make its own deviations manifest, thereby providing conditions for **recognition, learning, correction, and transformation**.*

*Within TLD itself, deviated information is proposed to appear only as an **obscuration of the state of integrity**: the original informational structure becomes less fully manifest, but matter has not yet formed. In four-dimensional spacetime, by contrast, the same underlying deviations become physically manifested through structures possessing curvature, twist, and physical interactions.*

*According to the geometric interpretation of IPTM, when deviation and curvature reach a critical regime, the informational structure may be conceptually represented as undergoing a transition away from its original orthogonal configuration. This transition may be compared illustratively with **Wick rotation**, which provides a mathematical image for* *relating different treatments of temporal and spatial coordinates.*

This comparison is strictly conceptual. IPTM does ***not*** *claim that Wick rotation as used in established physics is itself the physical mechanism through which matter is created.*

Karma as Geometric Memory

From this broader philosophical perspective, what is traditionally called ***karma*** *may be interpreted within the model as a form of* ***geometric memory of accumulated informational deviations****. When manifested within the physical world, these deviations contribute to patterns of attachment to four-dimensional phenomena and provide conditions for learning, transformation, and progressive restoration of informational integrity.*

The IPTM hypothesis therefore proposes a reversal of the conventional explanatory direction. Within established general relativity, the Einstein field equations relate spacetime curvature to the stress–energy content of the physical system. IPTM does not dispute that experimentally successful relationship within the physical domain. Instead, it proposes a deeper pre-physical stage in which ***informational curvature precedes both ordinary matter and physical spacetime curvature****.*

Accordingly, the model does not claim that observed spacetime curvature exists without the physical relationships described by general relativity. Rather, it hypothesizes that the physical curvature appearing in the Einstein description may itself be the manifested consequence of a more fundamental informational curvature that arose prior to matter.

The proposed hierarchy is therefore

Informational Deviation→Informational Curvature→Temporalization→Physical Spacetime→Fields and Matter
$I\text{dev}\rightarrow K\text{info}\rightarrow t_0\rightarrow t_1\rightarrow t_2\rightarrow(M, g\mu\nu)\rightarrow\{\Phi_{i,\text{Matter}}\}$

In this sense, ***matter does not create the primordial curvature proposed by IPTM; rather, matter and physical curvature are both later manifestations of a deeper informational deviation****.*

TLD is a theoretical proposal motivated by the possibility that our present description of reality may be incomplete and may therefore require a deeper foundation. n the geometrical interpretation of the model, when curvature and deviation reach a critical threshold, the informational structure may be visualized as undergoing a *geometric rupture or reorientation relative to its original orthogonal state. In the TLD model, the mathematical transformation known as Wick rotation is regarded as the formal expression of a deeper ontological mechanism involved in the transformation of information into matter. According to this hypothesis, primordial information does not enter physical spacetime directly, but first passes through an Interface Domain, where time emerges in successive degrees: from the timeless state of TLD, to proto-time t0, and then to minimal time t1, enabling the formation of dark matter. As informational inclination and torsion increase, the need to actualize information requires the formation of fully physical time t2, simultaneously with the emergence of photons, thereby completing the 4D world.*

Wick rotation is therefore understood as a mathematical indication of the transition between these temporal domains. In addition, the mechanism governing the circulation and transfer of information from TLD to 4D, as well as from 4D back to TLD, requires the presence of a filtering and discriminating operator: integral information is permitted to reintegrate, whereas deviated information is blocked and retained within 4D for correction. At the same time, this mechanism prevents information from entering the null space (kernel) of the matrix, where correction becomes impossible—a region that the model associates with Hell.

The transfer of information between the two domains is asymmetric: ***TLD knows all information within 4D, whereas 4D does not know the information of TLD***

TLD is a theory arising from an incomplete understanding of reality and still requires a foundation of empirical physical evidence. Evidence is needed both for the existence of TLD itself and for the proposition that time, space, and the Higgs field are "generated" through the processes proposed by TLD.
However, if TLD is regarded as a philosophical model or theory that is internally consistent with physics and reality, and if its mathematical structure can be deductively derived from the principles of TLD, then the HTTV theory, viewed from the perspective of the 4D world, may be constructed as follows:

A. FROM EMPTINESS TO TLD: THE TWO EMERGENCES OF FREEDOM

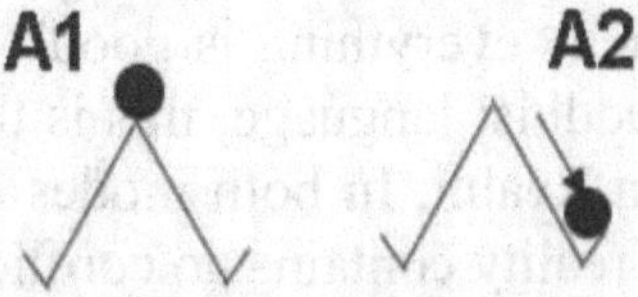

Figure 1.1A. Principle of Creation:
A1: Symmetric but unstable; the ball has high energy.
A2: Asymmetric but stable; the ball has low energy.

A familiar image in physics is the "Mexican-hat" potential. A ball placed exactly at the top represents a state of perfect symmetry. However, this state is unstable. It takes only a very small fluctuation arising from the principle of freedom—the fundamental principle of the creative universe—for the ball to roll away from the top and choose one among infinitely many possible directions. Symmetry still exists in the equations, but the actual state has spontaneously selected a particular direction. This phenomenon is called **spontaneous symmetry breaking**.
In classical metaphysical systems, it is often assumed that the world emerges from some absolute foundation: the Tao, Suchness, Emptiness, or Absolute Being. However, if that foundation is truly absolute, supersymmetric, non-dual, and undifferentiated, then the question immediately arises: why is there manifestation?
One possibility is to regard Emptiness not as "nothing," but as a state containing infinite possibilities for all forms of existence.

From this emerges TLD as the first pre-mathematical and pre-physical informational layer.
TLD has no space, no time, no energy, and no matter. It contains only pure structural informational relations.
It can be conceived as:
$\emptyset \rightarrow TLD$
This is the first "emergence of freedom."
According to the hypothesis, the **Timeless Domain (TLD/TLD)** is a **pre-geometric domain**, containing only **primordial, integral structural information**, with no matter, energy, or time in the physical sense.
In the primordial state of TLD, all informational structures are in a state of integrity, balance, and symmetry. Expressed in religious language, this is the condition in which "everything is good" in the Genesis account. Expressed in Buddhist language, this is the pure state of Suchness or the Dharma Realm. In both modes of expression, the original foundation of reality contains no conflict, separation, or opposition.
In the state of integrity, information exists as a unified whole, not yet divided into the opposing pairs of the dualistic world. Symmetrical structures do not cancel one another but coexist within a harmonious order. According to the interpretation of this model, such a state corresponds to the foundation of truth–goodness–beauty, or, more broadly, to universal ethical principles, which in philosophical and religious traditions are often described by different names such as Suchness, the Tao, God, or the Creator.
Integral information, Prajñā Wisdom, or Ultimate Knowing can also be understood in a profound ethical sense. All existence arises from the same single creative foundation; no entity exists as an absolutely independent essence. Therefore, at the deepest ontological level, there is no **separate self** that truly possesses its own existence.
That primordial phenomenon is the **Absolute Self**—the primordial, formless foundation that cannot be directly perceived through the senses. From this foundation, the entire phenomenal world is manifested as diverse forms of the same single reality. Although phenomena differ in form, they have no independent

self-nature; they are merely relative manifestations of that primordial foundation. *Because there is no time, structures and phenomena, although different, exist simultaneously in the same way, without before or after, without succession or arising-and-ceasing; one does not give rise to another, and although they are equal in status, they still remain distinct. If there is a distinction of before and after, such as the first day, the second day, the third day..., or above and below, it is because they are discovered and described by the observer or narrator in a sequential order of before and after, because* **the observer carries their own time**.

According to this view, it is precisely the manifestation of reality in the phenomenal world that reveals the **principle of non-self**. Because no phenomenon exists separately from the common foundation, there is also no independent and permanent self within phenomena. Therefore, non-self is not merely an ethical teaching or a method of practice, but also reflects the very ontological structure of reality.

In other words, the **Absolute Self**, although it cannot be directly perceived, remains the foundation of existence, **encompassing the totality of all things**. In different religious and philosophical traditions, this foundation is called by different names, such as the **Supreme Being**, the **Creator**, the **Tao**, **Suchness**, or **God**. These names do not necessarily refer to different entities, but may simply be different ways of expressing the same ultimate reality that transcends all symbols and language.

Unlike the **ego**, the **Self** has no time because it exists within TLD; all events and information appear simultaneously, with neither before nor after.

If all existence arises from the same foundation, then equality among all sentient beings is no longer merely an ethical norm, but becomes a natural consequence of the fundamental structure. From this arise universal ethical principles that the major religions express in different forms, such as the Ten Commandments in the Judeo-Christian tradition or the Five Precepts in Buddhism. In the TLD model, these principles may be regarded as manifestations of **integral information**, that is, informational structures that remain consistent with the primordial state of balance.

Another important characteristic of TLD is its **nearly absolute uniformity and symmetry**. Because it is not constrained by space, time, or matter, TLD also manifests the highest degree of freedom. It is precisely from this absolute freedom that the TLD model hypothesizes that some informational structures may naturally enter states deviating from perfect equilibrium. This is not a "violation" of a pre-existing law, but a natural consequence of freedom within an absolutely symmetric system.

. STRATIFICATION OF TLD INTO TWO DOMAINS: PRE-SPACE PRECEDES THE ARRIVAL OF BIASED, CURVED, AND TWISTED INFORMATIONAL STRUCTURES (ASYMMETRY), ALLOWING TIME TO EMERGE

If all structures in TLD were completely balanced, there would be no driving force for the emergence of the phenomenal world.

Therefore, TLD must allow structural freedom to exist.

This freedom leads to the possibility of configurations deviating from perfect equilibrium. Such deviation may be regarded as the loss of symmetry of the primordial state of Emptiness.

In the Language of TLD: Two Fundamental Layers of Information

Canonical Information — Non-string Information

This is the canonical informational layer of TLD. Its characteristics are: complete consistency; timelessness; invariability; and no need for correction. These structures may be associated with truth, Prajñā wisdom, and fundamental principles of harmony. They do not generate matter and do not directly participate in the process of physical manifestation.

Biased Information — String-like Information

When the deviation remains small, TLD still maintains its stability.

But when the deviation exceeds a critical threshold:

$\Delta I > \Delta I_c$

TLD begins to stratify. This is the second emergence of Freedom.

This is the dynamic informational layer. It is not an error in the ethical sense, but it has not yet reached a state of absolute equilibrium. These structures carry certain degrees of asymmetry

or deviation from the norm. For this reason, they require a process of correction and transformation.
This is the origin of the emergence of matter and the history of the universe.
Biased information is described by three fundamental quantities:

- (θ): the degree of informational inclination away from the state of integrity;
- (κ): informational curvature, hypothesized to be related to gravity;
- (τ): informational torsion, hypothesized to be related to particle spin and the degree of manifestation of material structures.
- **ρ-** density of information, ℓ "length of infornation related to the size

In addition, the scale of information, corresponding to the informational length ℓ and density ρ, consists of neutral characteristics that nevertheless influence its emergence into phenomena.
For example, biased information is the precursor of:

- **Photon:** θ minimally exceeds the threshold, with no curvature, $\kappa=0$, but with torsion, spin =1.
- **Higgs boson:** θ minimally exceeds the threshold, with curvature $\kappa>0$, no torsion $\tau=0$, scalar.
- **Hypothetical graviton:** θ minimally exceeds the threshold, with no curvature $\kappa=0$, and torsion, spin =2.
- **Neutrino:** θ is very small and just exceeds the threshold, $\kappa>0$, spin >0.

As the inclination angle θ increases, the difference between the existing state and the state of integrity becomes progressively greater. This is precisely what the TLD model calls **biased information**.
Biased information is not matter, nor is it yet energy. It still belongs to TLD and is not yet subject to time. However, deviation from equilibrium creates an intrinsic need: **the tendency to return to the original state of integrity**.
It is precisely this tendency that requires a new mechanism.

If all information appears instantaneously, then there can be no concept of ordering into before and after, arrangement, or correction, because correction always requires a sequence consisting of a before and an after. For a structure to move from a deviated state back toward a balanced state, a process must exist. And such a process can occur only when a mechanism of stratification and ordering emerges to restore the previously existing norms and order. **The mechanism that restores the established order is time itself.**

This is not merely a question of physics, but also one of philosophy, theology, and Buddhist thought. In the TLD model, time is not a primordial reality, but an emergent mechanism that arises in response to an intrinsic need within the informational structure itself.

From this perspective, time is not an independently existing entity. It appears only when there is a need for a **mechanism of informational ordering, organization, and stratification**, enabling the process of transformation from a biased state back toward the state of integrity. This process is initiated only when the degree of inclination exceeds a certain threshold, modeled by an index representing the degree of difference from the canonical state.

The model proposes that the process of **Information Pre-geometrization and Temporalization into Matter (HTTV)** takes place through three stages of time: t0, t1, and t2.

C. TIME AS A FACTOR OF ORDER

Time is not a component that exists independently from the beginning, but rather a **mechanism that allows information to reorganize itself**. In other words, time is a necessary condition for carrying out the self-correction process of informational structures.

When time begins to emerge, information no longer exists merely as before, as static relations within TLD. But at the beginning of this event and throughout the entire process above, there is still one point that needs to be emphasized. The first emergence of time within TLD is not yet physical time in the ordinary sense, but only **a potential for time**.

PRINCIPLE OF TEMPORAL EMERGENCE

In the TLD model, time does not exist as an independent entity. Time emerges only when information departs from its primordial symmetric state and needs to undergo a process of restoring order. Therefore, time is a manifestation of the process of restoring informational order, rather than the cause that creates that process. When symmetry is completely restored, the need for sequential ordering no longer exists, and time also loses its manifestation. **Within TLD there is a principle of restoring symmetry and order.**

At this stage, time does not yet have the capacity to actualize information into energy, mass, or matter. Its only role is to create the possibility of **stratifying, classifying, and ordering** informational structures according to their degrees of inclination and curvature. In other words, this is the stage at which information begins to have a "latent order," but there is not yet any significant effect of motion, no physical transformation, and no phenomenal world.

As the inclination angle θ increases, the curvature of information also increases accordingly. Informational structures are therefore gradually distributed into different levels, from near to far relative to the primordial state of integrity of TLD. The more inclined a structure is, the farther it is from the original pure state and the more strongly it manifests the need for self-correction.

This primordial level of time may be called **proto-time**, denoted by t_0. Proto-time is not a continuous flow of time as we experience it in the four-dimensional world. It is only a **pre-temporal principle**, expressing the capacity to establish order among informational structures. Proto-time itself opens the way for the emergence of higher levels of time, from **minimal time** t1 in the Interface Domain to **full time** t2 of the material world.

According to the TLDmodel, proto-time is therefore not a dimension of spacetime, but the **first transitional step** from a completely timeless reality toward a reality capable of generating time. It marks the point at which TLD begins to differentiate, while still not departing from its non-material nature.

Next, time truly emerges. Under the influence of time, information gradually becomes actualized into energy, physical fields, mass, elementary particles, matter, and finally biological systems capable of cognition. The entire four-dimensional world becomes a dynamic process of information.
According to this view, matter is not the ultimate goal of creation. **The materialization of information is only a means of carrying out the process of self-correction.** The material world becomes an environment in which informational structures can be concretely manifested, can interact, choose, change, and gradually restore the primordial state of equilibrium. The manifestation of matter is not regarded as an end in itself, but as a consequence of the need for information to reveal its own state, thereby creating the conditions for the process of learning, adjustment, and transformation.
This view provides a new understanding of the meaning of the universe. The universe does not exist merely to contain more galaxies, planets, or living beings. The emergence of matter is a necessary stage in the process of restoring informational order. In this sense, the history of the universe is also the history of the process of self-adjustment of informational structures.
If biased information is regarded as the source of imbalance, then the evolutionary process of the universe can be understood as a movement toward restoring the state of integrity. This is an interesting point of convergence between physics, ethics, and spirituality. In the phenomenal world, the purification of the ego and the development of morality and wisdom not only carry humanistic meaning but may also reflect the natural direction of the process of restoring informational order.
As informational structures gradually return to the state of purity, the role of time also gradually diminishes. If the entire informational structure regains its primordial equilibrium, the need for the corrective process no longer exists. At that point, time completes its mission and withdraws together with the disappearance of motion, energy, and matter. Reality returns to the original timeless state of TLD—a domain of purity, integrity, and neither arising nor ceasing.

According to the TLD hypothesis, this is not the end of the universe but the completion of a cycle of informational actualization: from purity, through differentiation, to self-correction, and finally back to purity.

D. BIASED INFORMATION ACTIVATES THE EMERGENCE OF TIME

1. MINIMAL BIASED INFORMATION

At the t_0 (proto-time) stage, the degree of inclination remains small and still lies within limits that can be accepted relative to the canonical state. At this point, time serves only as the capacity to order, classify, and record information. There is no energy, no physical field, and no formation of matter.

2. BIASED INFORMATION AS THE PRECONDITION FOR MATTER

When Informational Inclination Exceeds the Threshold: TIME t1

When informational inclination exceeds a critical threshold, the **need for correction and restoration** becomes necessary so that the deviated informational structure can continue to be adjusted and perfected within TLD. The first condition of this corrective process is that the deviation must become **manifest and capable of being recognized**. For this to occur, biased information must undergo a process of increasingly explicit manifestation and ultimately become **materialized**—that is, actualized in the form of matter.

In the model proposed here, the tendency toward restoration is regarded as an intrinsic characteristic of the order of TLD. A deviation cannot be corrected if it remains completely latent; first, it must be made sufficiently explicit so that the nature and consequences of that deviation itself can be recognized.

Therefore, the model proposes two levels or two branches of manifestation through which biased information becomes increasingly explicit and capable of being corrected. **Time emerges as an imprint of this process of making deviation manifest.** In this sense, time initially is not the cause of

transformation, but rather **the sign of a deeper process in which biased information is made manifest, ordered, and becomes an object capable of being recognized for correction.**
This principle can be summarized as follows:

Informational Deviation⟶Need for Manifestation⟶Temporalization⟶Deviation Becomes Manifest⟶Correction and Restoration

a. When the inclination exceeds the minimum threshold, the need for adjustment to return information to the state of integrity becomes more urgent, with an intensity that increases according to the degree of deviation. This is the t1 (minimal time) stage, corresponding to the formation of the Interface Domain, where information begins to manifest in the form of curvature and pre-material effects. In this model, dark matter is hypothesized to belong to this stage. Force (pre-energy) is minimal but sufficient to activate the occurrence of an event.

b. When the inclination continues to increase
At the t_2 (physical time) stage, time becomes the complete physical time of four-dimensional spacetime. Here, quantum fields, elementary particles, and the entire material world as we observe it emerge.
Within TLD, biased information manifests only as an obscuration of the state of Suchness, that is, reducing the capacity of primordial information to become manifest without yet forming matter. In contrast, within four-dimensional spacetime, the same inclination is actualized in the form of material structures carrying curvature, torsion, and physical interactions.
According to the geometric interpretation of the model, when curvature and inclination reach the critical limit, the informational structure, after time has been added to it, can be conceived as a portion of biased information being broken away from its original orthogonal state. This image is illustratively associated with **Wick rotation**, as a mathematical symbol for the transition from the pre-geometric domain to the material domain. This event is also necessary because information, once it has time, cannot remain within TLD. However, this is only a conceptual comparison, suggestive in its logic insofar as it separates TLD from the domain

possessing time, but it cannot yet be asserted that Wick rotation in physics is itself the mechanism that creates matter.

From this perspective, what is commonly called **karma** may be understood as the geometric memory of accumulated inclinations within the informational structure. When manifested in the material world, these deviations contribute to the tendency to become attached to the phenomena of the four-dimensional world and become a driving force for the process of learning, transformation, and perfection of each individual.

3. TIME t_2 EMERGES TOGETHER WITH THE PHOTON

Full time and electromagnetic light (the photon) are two different phenomena, but they are closely related. When the degree of biased information exceeds a critical threshold, according to the TLD hypothesis, time and the photon begin to emerge in the transition from TLD to the physical world.

The reason **time t2 and the photon emerge in close association with one another** is that, in the TLD model, the function of time is to create a sequence so that informational deviations can be **manifested, distinguished, and gradually corrected**. However, a deviation can be recognized only when it is manifested. The photon, as the quantum of the electromagnetic field and a means of transmitting physical information, contributes to making those differences capable of being manifested and recognized in the physical world.

In this sense, **time creates the conditions for the photon to be actualized, while the photon contributes to making the progression of time observable and measurable**. Time is not the photon, and the photon is not time; they are two different but complementary components in the process of actualization: **time establishes the before–after order, while the photon enables the differences within that order to be manifested and recognized.**

E. PROPOSED THEORY OF INFORMATION PREGEOMETRIZATION AND TEMPORALIZATION INTO MATTER (IGTM)

(Proposed Theory of Information Pregeometrization and Temporalization into Matter (IGTM))

1. PRINCIPLE

"Matter is not the starting point of reality, but the point at which structural deviation takes form. Life is not a biological accident, but an engine of consciousness designed to resolve informational paradoxes."

Throughout the history of physics and philosophy, we have always been confronted with the question of the origin of matter and time. Einstein's General Theory of Relativity teaches us how matter curves spacetime, but cannot answer why matter exists in the first place. Quantum physics describes the world in terms of information, but leaves a large gap between abstract information and tangible reality.

Using the language of Differential Geometry and Systems Theory, this book proposes that ultimate reality is an absolutely quiescent network of information. When deviations arise, they are not destroyed, but are enclosed within self-curving and self-twisting informational strings. This twisting generates intrinsic time, gives rise to mass, and manifests as the 4D physical world that humans perceive. Matter, therefore, is both a structural debt (Karma) and, at the same time, a safety net, an ultimate experimental field upon which life can rely, reshape, and correct itself before returning to its pure source.

This is a nonlinear geometric theory intended to describe the nature of information in the **Timeless Pre-Space (Timeless Domain – TLD)**. According to this hypothesis, when information remains in its canonical state within TLD, geometry, matter, and time in the physical sense do not yet exist. However, when a pregeometric inclination characterized by the angle θ appears, due to the breaking of primordial symmetry, curvature κ and torsion τ also begin to form simultaneously. This theory is not developed from an 11-dimensional spatial theory or from String Theory. It develops as a necessary requirement when TLD is conceived as the foundation of 4D reality.

When the degree of inclination remains below the threshold required for manifestation, information remains in the nonphysical state of TLD. When the degree of deviation exceeds a critical threshold, the model hypothesizes that a pre-temporal ordering mechanism begins to emerge. This mechanism is called **proto-time**, denoted by t_0.

Proto-time is not physical time and is not a dimension of spacetime. It merely represents the capacity to establish the ordering, stratification, and before–after relationships necessary for the process of informational restructuring.

From the proto-time t0 foundation, HTTV proposes two different branches of temporalization:

$t_0\{\longrightarrow t_1 \longrightarrow S_{DM}, \longrightarrow t_2 \longrightarrow S_{4D}.$

In this formulation, t_1, or **minimal time**, is hypothesized to be associated with the Interface Domain and the dark sector; whereas t_2, or **physical time**, corresponds to the complete temporal regime of 4D spacetime, where quantum fields and ordinary matter can manifest.

S_{DM}=Interface / dark sector,S_{4D}=4D physical sector.

Importantly, HTTV **does not assume a compulsory process**. Instead, t_1 and t_2 are regarded as two different modes of temporalization sharing the same proto-time foundation t0. Therefore, dark matter in the t_1 branch is not automatically regarded as a precursor that must transform into ordinary matter in the t_2 branch.

HTTV is not constructed as a consequence of String Theory or an 11-dimensional spacetime model. It is proposed independently from the TLD hypothesis. Possible similarities with String Theory, M-theory, Hilbert space, or gauge structures will be examined later as possible mathematical relationships, rather than being used as premises to prove the existence of TLD.

The three quantities θ,κ,τ do not exist independently but interact with and influence one another through a nonlinear mechanism. When the degree of inclination exceeds a critical threshold, the informational system transitions from the state of integrity into a **pregeometric state**, creating conditions for the formation of primitive geometric structures.

The **IGTM** theory is proposed to geometrize primordial structural information and describe the process by which information is gradually transformed into matter. According to this model, the

pregeometric state activates a mechanism of informational stratification and self-adjustment through progressive stages of time: from **proto-time** t_0, to **minimal time** t_1, and finally to **physical time** t_2.

In this theory, **time is not regarded as a fundamental entity of the universe, but as an organizing factor that emerges when information needs to be arranged, maintained in order, and restructured toward actualization**. This conception arises from the hypothesis that time has a nature independent of space and **emerges only when informational structure requires a mechanism of sequential ordering**. In other words, time does not exist because matter already exists, but emerges as a consequence of the need to organize and transform information.

This process leads to the stratification of information within TLD. At the **proto-time** t_0 stage, time does not yet have physical meaning but merely represents the **capacity to arrange and maintain the order** of information. In this state, structural deviations still remain within minimally acceptable limits, so TLD (or TLD) remains in an almost pristine state. At that point, the need to correct or restructure information is not yet great enough to alter the overall structure of TLD.

However, when the degree of informational inclination exceeds a critical threshold, correcting and reorganizing the deviated structures becomes an inevitable requirement for maintaining the consistency of TLD. This need activates the transition from **proto-time** t_0 to **minimal time** t_1. Time at the t_1 stage remains very primitive and is not yet physical time in the ordinary sense, but it is sufficient to allow processes of stratification, adjustment, and informational transformation to begin.

At the t_1 stage, dark matter (DM) is hypothesized to be the first form of matter appearing within the **Interface Domain**, where information already possesses gravitational curvature but does not yet fully manifest the characteristics of ordinary matter. When time reaches the t_2 stage, the first quantum structures of visible matter begin to appear, while physical quantities such as **gravity, mass**, and **spin** gradually become observable properties in the 4D world.

2. THE THREE PRIMORDIAL CODES: Angle θ, Curvature (κ), and Entropy/Spin (τ) Activate Time t

If matter has not yet arisen within TLD, then all things are shaped according to purely structural mathematical parameters called the **Three Information Precursors**. These three codes do not arise spontaneously, but are latent phenomena activated by biased information; this issue will be explained more clearly in later chapters. All three parameters are both independent and mutually dependent as follows:

Θ is primordial, representing the beginning of inclination. S may manifest even when there is no inclination angle θ, because both are manifestations occurring when symmetry collapses. Nevertheless, inclination allows τ to develop more easily. Curvature occurs only when there is inclination >0; conversely, torsion makes curvature more difficult to manifest according to the equation:

$\kappa=\alpha 1+e^{-\tau\theta}$.

In summary:

(1)M(θ,κ,τ)θκ=F(θ,τ)τ=F(θ,κ)(ethical deviation)(weight or Karm a)(Entropy/Spin τ) (2)ℓ,ρ:scale and density of information.

i. Phase angle θ (Precursor of Time and Frequency): Within the TLD domain, θ is a fixed complex phase value, functioning as a geometric key code. It contains the algorithm governing how a system will operate. When passing through the Interface Region and undergoing Wick rotation, this static phase angle θ is forced to vary relative to the real time axis t, giving rise to oscillation with physical angular frequency ω, thereby creating the cycle and evolution of the particle.

2. Curvature number κ (Precursor of Space and Momentum): The curvature number κ in TLD determines the geometric structure, distribution density, and topological distance of informational nodes. It is the code determining how space will curve or expand when transformed into 4D reality. When interacting at the boundary region, it transforms into a geometric momentum vector, [h/2π; needed to indicate the Planck scale of measurement], forcing physical spacetime to form particle folds.

Entropy/Spin τ (Precursor of the Internal State): τ is the code that determines the internal symmetry structure and the degree of mathematical freedom of the data stream. In TLD, τ determines invariant topological properties (topological invariants). When the τ code is decoded through the Interface Region, it manifests in 4D space as particle spin (determining whether the particle is a force-carrying boson or a matter-forming fermion) and the law of increasing entropy in thermodynamics. It may be separated into two factors: τ **(or σ) perpendicular to the surface and τ (tau) parallel to the surface.**

The inclination that activates the manifestation of the codes may be regarded as an axiom of the model and currently remains an unknown. However, it may be assumed that the value of this unknown is small and lies within an acceptable error limit. For example, the 5% threshold is commonly used in experimental statistics as a conventional level for evaluating the significance of results.

When the inclination θ increases, κ and τ, together with the density and magnitude of information, form a force within TLD/the Vacuum, creating a high degree of symmetry breaking that exceeds the intrinsic vacuum energy of the VEV, allowing a minimum force to escape the constraints of the vacuum. At this point, it is not sufficient to create weight, but may be sufficient to propagate and move a particle at velocity c, with spin due to that inclination θ; that particle is the photon.

3. FROM INFORMATIONAL GEOMETRY TO MATTER: THE ROLE OF $\theta,\kappa,\tau,\ell,\rho$ IN HTTM THEORY

a) GEOMETRIC CHARACTERISTICS OF THE OSCILLATION MODE

To package the three precursor codes θ,κ,τ in the timeless TLD domain into a rigorous mathematical structure as the Original Informational Wave Function.

In the TLD model, primordial structural information is not yet matter, nor does it carry energy in the ordinary physical sense, because it exists within the timeless domain. An informational structure can be preliminarily characterized by five parameters:

$I=(\theta,\kappa,\tau,\ell,\rho)$

where θ is **informational deviation**, κ is **informational curvature**, τ is **informational torsion**, and ℓ is the **characteristic structural informational length**. These five quantities are not necessarily completely independent. It may be hypothesized that primordial inclination generates or modifies the geometry of the structure, for example:

$\kappa=f(\theta,\tau)$.

According to this view, θ does not directly become mass or energy. It represents the degree to which the structure has departed from its original state of integrity and symmetry. When

$\theta\leq\theta c$,

the structure remains within the timeless domain of TLD and has not yet manifested as a physical excitation. But when

$\theta>\theta c$,

an ontological threshold is crossed: the mechanism of temporalization begins to operate. Time does not create information from nothing, but provides the condition under which a previously static and timeless structure can be **sequentialized, oscillated, and actualized**.

It may be figuratively conceived that **time is "pumped" into the informational structure**. This expression does not mean that time is a substance or physical energy that is mechanically pumped. It describes an ontological phase transition: when temporalization emerges, the structure $(\theta,\kappa,\tau,\ell,\rho)$ begins to acquire the capacity to manifest in the form of an intrinsic frequency ωI.

Hypothetically, this frequency may be represented by a general function:

$\omega I=\Omega(\kappa,\tau,\ell,\rho),\theta>\theta c$.

Here, κ,τ determine the intrinsic geometry of the structural information, while ℓ establishes the size scale of the structure. A structure with a smaller ℓ may correspond to a higher-frequency mode, while changes in curvature and torsion alter how that structure oscillates and manifests.

Once temporalization has occurred, frequency can be connected to physical energy through the familiar quantum relation:

$E=\hbar\omega$.

Therefore, TLD proposes a sequence of actualization:

Biased I⟶θ⟶t⟶physical manifestation.

The photon is the simplest boundary case illustrating this hypothesis. Within TLD, one may set

$\kappa_\gamma=0,\tau_\gamma=0,\ell_\gamma>0,m_{0\gamma}=0$.

The photon therefore has no rest mass,

$m_0=0$,

but still possesses spin, frequency, and energy. Structural length is hypothesized to be a precursor proportional to the observed wavelength:

$\ell_{\gamma TLD}\propto\lambda_{\gamma 4D}$,

where ∝ means proportional to, and λ_γ is the wavelength of the photon.

Therefore, photons may be **identical in nature but different in state**. Infrared and ultraviolet photons both have

$m_0=0,s=1$,

but possess different structural scales:

$\ell_{UV}<\ell_{IR}$,

corresponding in 4D to

$\lambda_{UV}<\lambda_{IR}$,

and therefore

$\omega_{UV}>\omega_{IR},E_{UV}>E_{IR}$,

where ω is frequency.

Thus, in the TLD interpretation, differences in photon color, wavelength, and energy do not necessarily begin in the 4D world; they may reflect differences already existing within the precursor structural-information configurations. In other words, **different phenomena originate from different structural natures**. However, these differences in nature often cannot be directly recognized at the 11D/4D level, but become apparent only when considering the deeper foundation of—

The neutrino may be regarded as the next step in this process. If the photon lies at the boundary,

$\kappa_\gamma=0$,

then the neutrino is hypothesized to have

$0<\kappa_\nu\ll 1$,

that is, a very small informational curvature that is not completely zero. When temporalization actualizes that structure, a very small rest mass begins to appear:

$m_\nu > 0, m_\nu \ll m_e$.

According to this hypothesis, as informational curvature increases, the structure becomes increasingly capable of localization and manifestation in the form of rest mass. Electrons, muons, hadrons, and other massive particles may be regarded as more complex configurations in which (κ,τ,ℓ) combine in different modes. However, this should not be understood simply as meaning that the mass of every particle is proportional only to κ; composite particles such as the proton also contain very large internal interaction energies. TLD proposes only that these physical characteristics must have corresponding precursor structures.

From this, a broader principle may be stated:

All physical manifestation has a structural precursor in the TLD.

Every physical manifestation is hypothesized to have a structural precursor in TLD.

Mass, spin, wavelength, frequency, and quantum states are not necessarily created from nothing when the 4D world emerges. They may be different forms of manifestation of a deeper informational structure already encoded by θ,κ,τ.

Accordingly, time is not the author that creates structure, but rather the **condition that allows structure to become phenomenon**. Before temporalization, information *exists as structure*; after temporalization, structure can *unfold as a process*. It is precisely this transition from **being** to **becoming**, from timeless structure to phenomena possessing frequency, energy, and history, that TLD hypothesizes lies at the foundation of the emergence of matter:

$(\theta,\kappa,\tau,\ell)_{TLD} \longrightarrow \omega_I \longrightarrow E_I \longrightarrow Matter_{4D}$.

In this sense, matter is not the opposite of information. **Matter is a form that structural information can assume when timeless structure is brought into the order of time.**

Time and the photon are two different phenomena, but within the TLD model they may be regarded as **arising together in the process of actualizing biased information**, with different but complementary functions. Time creates the necessary before–after order for a process to occur, while the photon allows differences within the physical world to be transmitted, to interact,

and to be recognized. In this sense, time and the photon together contribute to making **biased information manifest**, thereby creating the conditions for the process of correction and restoration of order.
This emergence does not necessarily have to be understood as an influence coming from outside TLD. In the TLD hypothesis, **the tendency to restore order is an intrinsic characteristic of TLD itself**. When information remains in a state of integrity, there is no need for correction and no need for a before–after process. But when structural freedom permits the emergence of informational deviation θ, a difference between the existing structure and the state of integrity begins to appear. When the deviation exceeds a critical threshold,
$\theta > \theta c$,
temporalization begins. Time is therefore not the primary cause that creates the deviation; it is the **mechanism that allows the deviation to be stratified, recognized, and sequentialized**, thereby making the restoration of order possible.
The photon may emerge near this transitional threshold. In the TLD interpretation, the photon is one of the earliest physical manifestations of structural information after temporalization. With zero rest mass and velocity c in vacuum, the photon can transmit differences from one region of the physical world to another. Thus, if time creates the **order of change**, the photon provides one of the fundamental means through which **change is communicated and becomes observable**.
This idea may be represented generally as:
Biased I⟶θ⟶t⟶photon ordering
(information I inclined by the angle θc will give rise to a photon from the TLD vacuum).
According to this view, the emergence of time and the photon is not composed of two entirely accidental events standing beside one another. They are two different functions of the same process of transition from **timeless information to the phenomenal world**: time allows difference to unfold, while the photon allows difference to be transmitted and manifested.
At a deeper level, TLD may hypothesize that the capacity for informational deviation θ to arise is a primordial property of the

structure of the universe. The threshold value θc may function as a **structural constant of TLD**, similar in spirit—but not identical in physical nature

4. PROPOSED EQUATION FOR THE OSCILLATION FREQUENCY OF THE INFORMATIONAL STRING

The vibration of an informational string is hypothesized to depend on four fundamental parameters: the inclination angle θ, curvature κ, torsion τ, and structural length ℓ. However, because the influence of θ is already contained within κ and τ, it is not necessary to include θ directly in the oscillation equation.

On this basis, an equation may be proposed to describe the **intrinsic angular frequency of the informational string**, denoted by ωI, in which the two geometric components κ and τ are regarded as two orthogonal axes contributing to the resultant vibration, while the angular frequency ωI is hypothesized to be inversely proportional to the structural length ℓ. Thus, an informational structure with a shorter length may correspond to a higher oscillation frequency, while κ and τ determine it.

$$\omega_I = \frac{\omega_0 \ell_0^2 \sin\theta}{\ell} \sqrt{a_\kappa \kappa^2 + a_\tau \tau^2} + \frac{\partial \theta_n}{\partial t}$$

In the above formula, in addition to κ and τ, the informational inclination angle θ also plays a particularly important role. In the HTTV/TLD model, θ is hypothesized to be the parameter that initiates the temporalization process, leading to the emergence of physical time t_2 and the photon.

Meanwhile, the structural length ℓ is hypothesized to be the principal parameter governing the oscillation frequency of the informational string. Therefore, photons with different wavelengths, such as ultraviolet (UV) and infrared photons, may be associated with different values of the structural length ℓ in TLD.

The oscillation amplitude represents the probability density for the emergence of matter (elementary particles) at the interface. It obeys a forced wave equation whose motion is

governed by the geometric structure, varying harmonically with time at angular frequency ω (represented by the term eiωt).
Left-hand side: Describes the wave propagation of the quantity X(s,t) through space s and time t with velocity v.
Right-hand side: Represents the **driving force** acting on the system. The amplitude of this force depends on the geometric and physical components.
Suy ra biên độ giao động:

$$\left(\left(\frac{\partial^2}{\partial t^2} - v^2\frac{\partial^2}{\partial s^2}\right)X(s,t) = \boldsymbol{F}\mathit{cuõng\ bức}$$

$$= \mathcal{G}11D \cdot \kappa \cdot \cos\theta \cdot \boldsymbol{N} + \tau \cdot \sin\theta \cdot \boldsymbol{Bei\omega t}\right)$$

$$P(A(s) \propto \frac{\mathcal{G}_{11D}}{\omega^2 - \omega_n^2} \cdot \sqrt{\kappa^2\cos^2\theta + \tau^2\sin^2\theta)}$$

$\mathcal{G}_{11D}$: The coupling coefficient of the metric field from the 11D bulk space down to the brane world.
At this interdimensional boundary, the equation of state of reality is determined through the geometric Frenet–Serret relations. When inclined quantum information generates a phase angle θ, it activates the **Berry geometric effective potential** through the combination of extrinsic curvature κ and spatial torsion τ. These quantities not only direct the flow of energy moving orthogonally from the interior of the 4D QFT brane toward the higher-dimensional 11D space, but also directly restructure the oscillation frequency ωn.
The oscillation amplitude A(s) no longer depends on macroscopic mechanical collisions, but is the result of a structural resolution function: geometric symmetry breaking (κ,τ), combined with phase inclination (θ), forces the vacuum to undergo spontaneous symmetry breaking, compelling the interdimensional strings to oscillate and thereby manifesting pure information as energy particles of the material world.
Frequency and Amplitude (Photon Wave Equation)
Substituting κ=0 into the amplitude equation:

$$(A(s) \propto \frac{\mathcal{G}_{11D}}{\omega^2 - \omega_n^2} \cdot \sqrt{0^2 \cdot \cos^2\theta + \tau^2\sin^2\theta} = \frac{\mathcal{G}_{11D} \cdot \tau|\sin\theta|}{\omega^2 - \omega_n^2})$$

the denominator ω2−ωn2 indicates that the closer the driving frequency ω is to the natural frequency ωn, the smaller the denominator approaches zero, causing the oscillation amplitude A to become extremely large.

1. EVOLUTIONARY STATE OF THE ENERGY OF THE VIBRATING INFORMATIONAL STRING

In HTTV/TLD, we hypothesize that the precursor configuration of a vibrational mode is characterized by $(\theta,\kappa,\tau,\ell)$. When formally compared with String Theory, the vibrational modes of strings provide a suggestive mathematical model for how structural information may be actualized into different energy states after temporalization.

Further Reading

In quantum mechanics:

$i\hbar\partial t\partial|\Psi\rangle=H|\Psi\rangle$,

where the Hamiltonian H is the operator that generates time evolution. If the state is an eigenstate of H:

$H|\Psi\rangle=E|\Psi\rangle$,

then:

$E=\hbar\omega$.

Therefore, in the TLD model, after temporalization, we may write:

$HI|\Psi I\rangle=\hbar\omega I|\Psi I\rangle$

to connect the equation for ωI with the Hamiltonian.

The string and its vibrational characteristics are controlled by information possessing deviation θ, curvature κ, and torsion τ within TLD. The vibrational mode of the string may be represented by the function:

$\omega=\left(e^{\left\{-\frac{iHt}{\hbar}\right\}}\right)$. In which the operator H (Hamiltonian) provides the detailed mathematical representation based on

$\omega=\Omega(\theta,\kappa,\tau)$

and governs the time evolution of the system $\omega=\left(e^{\left\{-\frac{iHt}{\hbar}\right\}}\right)$ ta có

$H_I=\frac{1}{2m}\left(\frac{-i\hbar\delta}{\delta x}-\hbar\tau\mathbf{J}\right)^2-\frac{\hbar^2}{4m}\kappa^{-2}+V_\kappa(\kappa,\ell)$

$P_s=-i\hbar\delta/\delta x$: đong lực độ xoắn.

T: độ xoán, $\acute{A}=\hbar\tau\mathbf{J}(\theta)$,

J(θ) : toán tử quay, $\begin{pmatrix} 0 & -e^{-i\theta} \\ e^{i\theta} & 0 \end{pmatrix}$

$(-i\hbar\delta/\delta x - \hbar\tau\mathbf{J})^2$ thanh phan xoán

6. EQUATION OF THE STATE OF OBSCURATION OF SUCHNESS IN THE HTTM MODEL

Biased, curved, and twisted information obscures the nature of Suchness in TLD and forms the veil of ignorance in the 4D world.

In contemporary particle physics, the Standard Model is based on the gauge group:

SU(3)×SU(2)×U(1).

These three groups describe deep types of symmetry in nature.

i. U(1)

The U(1) group is associated with electromagnetism. It has one generator, corresponding to the photon:

U(1)→γ.

In the language of TLD, U(1) may be regarded as the simplest phase symmetry of the informational fiber. An internal phase changes, while the informational standard is preserved.

From an ethical perspective, the photon is a manifestation of **Delusion/Ignorance (Si)** within the Three Poisons of **Greed, Anger, and Delusion (Tham–Sân–Si)**, because it obscures and replaces the primordial light of wisdom.

ii. SU(2)

The number of generators of SU(n) is:

n^2-1.

The SU(2) group is associated with the weak force. It has three generators,

$2^2-1=3$,

corresponding to the three weak bosons:

W+,W−,Z0.

Therefore:

SU(2)→W+,W−,Z0.

The weak force is distinctive because it is associated with left–right asymmetry, or parity violation. In the language of TLD, this may be regarded as a trace of chiral bias in the informational structure as it enters 4D. In ethics, it represents **Anger (Sân)**.

iii. SU(3)

The SU(3) group is associated with the strong force.
It has eight generators, corresponding to eight gluons:
$3^2-1=8$,
and therefore:
SU(3)→8 gluons.
The three “colors” of quarks are not physical colors, but three internal states. In TLD, they may be understood as three modes of internal transformation of the informational fiber after it has entered the material domain.
Thus, the entire Standard Model may be regarded as a system of symmetries of informational manifestations after they have entered 4D.
In other words, **Greed, Anger, and Delusion (Tham–Sân–Si)** are not merely psychological states. In TLD/TLD, they may be regarded as forms of information that have deviated from their original integrity.

- **Ignorance (VM)** appears in the form of the deviation angle θ. It activates the 12 toxic channels of **Greed–Anger–Delusion (Tham–Sân–Si)** within the Standard

$$\left(\theta^a T_a = \theta_{U(1)} Y + \sum_{b=1}^{3} \theta^b_{SU(2)} T_b + \sum_{c=1}^{8} \theta^c_{SU(3)} \lambda_c\right)$$

Model symmetry structure, indicating the types of particles and interactions that produce obscuration. Y(1), T, and the rotation angle θ are the originating elements that generate the generators in Yang–Mills theory.

This activation bends and twists the perceptual membrane, generating Karmic energy (κ^2) and Spin particles (τ^2), producing the total **Obscuration Density** (M) that isolates the practitioner within the 4D realm.

$$\mathcal{M}(\theta,k,s) = \mathrm{Tr}\left(k_{\mu\nu}k^{\mu\nu}\right) + \alpha\mathrm{Tr}\left(s_{\mu\nu}s^{\mu\nu}\right)$$

- **Energy** is the square of Torsion.

In terms of dynamics, the kinetic-energy term associated with α affects the quantum loop, generating the obscuration effect represented by Tr. The **Trace**, or Tr, is the diagonal trace of the matrix running from top to bottom and from left to right, representing the rows and columns of the numerical matrix that expresses the veil of ignorance.

Activation of the "detoxification" operator U−1: U−1 represents the process of restoration or reversal of the transformation; Tr is used to measure an invariant aggregate quantity of the structure in order to cleanse the entire twisted and obscured structure.

$(\mathrm{Tr}(M_{\mathrm{new}}) = \mathrm{Tr}(UM_{\mathrm{chaotic}}U^{-1})$ vì $UU^{-1} = I)$

To bring these values back to zero and restore the **Nature of Suchness**, the process will be carried out by a projector or a separate restoration operator.

7. WICK ROTATION

The Dimensional Bridge: Wick Rotation and the Birth of Real Time

How can a static, timeless informational reality in TLD transform into a dynamic universe in 4D space? The answer lies in the ultimate geometric mechanism at the boundary of reality: **Wick Rotation**.

At the ontological level of reality, TLD and the physical 4D universe do not lie sequentially adjacent to one another, but exist in an **orthogonal** relationship. Wick Rotation plays the role of a **dimension-shifting operator**, performing a 90∘90^\circ geometric rotation in the complex plane, transforming the imaginary time axis τ\tau of TLD into the real time axis tt of the dynamical world:

4. t=−iτ

Further Reading

5. When this rotation is activated at the boundary, the static informational distribution function of TLD is immediately pushed into a quantum dynamical phase:

6. Informational deviation makes the system dynamic, represented by HH (the Hamiltonian), which is the operator of the total energy of the system and has the function of indicating the energy/degree of manifestation of the quantum system upon which it acts.

7. e−iHt/ℏe^{-iHt/\hbar}

8. transforms imaginary time τ\tau into tt, bringing the system to life.
9. **The equation above and e−iHt/ℏe^{-iHt/\hbar} are precisely the bridge that makes the strings in String Theory move, giving rise to elementary particles.**
10. This transformation converts the static phase constant θ\theta in TLD into an angular frequency varying with real time, while simultaneously releasing the wave vector κ\kappa into physical momentum. Real time is born not from a mechanical explosion of matter, but from the very moment when the complex coordinate axis of TLD "rotates" through Wick rotation.

11. 1. THE MATHEMATICS OF THE INTERFACE DOMAIN REFLECTS THE ARCHITECTURE OF REALITY

12. "After defining the mode and vibrational energy of the materialized informational string, the following section studies the role of the Interface in: the process of classifying information within the Interface; TLD → Interface (with t_1) and 4D (four-dimensional space) with t_2; the reverse direction, Interface 4D → TLD; the information-filtering network; and the processing of information that cannot be corrected by pushing it into the space of Hell."

a) The Origin of Structure and Orthogonal Decomposition

Every structured reality requires an absolute point of departure—a reference system that is not distorted by the flow of time or the disorder of matter. In this architectural model, that point of origin is the **Timeless Domain (TLD)**, mathematically denoted by the origin t0. From the perspective of quantum physics, t0 is not a static spatial location, but a **pure original state vector**, existing within an infinite-dimensional function space. At this foundational anchor point, the system reaches a state of absolute order with **Von Neumann Entropy equal to zero** (using a matrix to measure the displacement of information when compared with the canonical state). There is no passage of time, no material particle, and no geometric distance; all reality exists only in the form of pure informational potentials.

However, in order for information to manifest and diagnose itself, the system performs a great branching step directly at the t_0 gateway through the principle of **Orthogonal Decomposition**. By applying two independent linear operators, reality divides into two worlds: P1 corresponds to t1, continuing in the same direction as TLD, while P2 corresponds to t2, orthogonal to TLD.

$$P_1)\, và\, (P_2)\, tácđộngtrựctiếpvàovectơgốc\, (|\Psi_0\rangle),$$

The P1 branch weaves the **Interface Space t1**—a linear, neutral informational plane where the operational laws are uniform in order to establish a diagnostic error-filtering membrane.

$$|\Psi_0\rangle \quad P_1|\Psi_0\rangle + P_2|\Psi_0\rangle$$

In contrast, the P2 branch projects directly into **4D Space t2**—a nonlinear dynamical domain where information coils and twists itself in order to become materialized as matter. To ensure that these two worlds are completely structurally independent and do not interfere with one another, the two operators must satisfy the condition of absolute orthogonality.

From the anchor point **t0**, the journey of information officially begins to enter specific geometric spaces.

i. Interface Space t1 and the Geometry of Purification

The **Interface Space t1** is woven from a network of informational strings together with the property of absolute flatness. The property of "flatness" here does not merely mean ordinary Euclidean geometric flatness, but represents the **linearity and consistency of the laws (Linearity)**. On the **t1** plane, informational strings operate in parallel or intersect according to strict rules of order without becoming entangled, curved, or forming unintended data knots.

13. This is a neutral transmission environment that does not itself generate gravity or attractive forces that deflect the data entering from TLD. Within the Interface, biased information will become manifest through the mechanism of a matrix that takes the residual after subtracting the integral information (which lies along the linear path).

It is precisely because of this flat and quiescent property that $t_1$functions as a **diagnostic fluorescent screen**. When an informational state vector carrying deviations from reality returns

to the Interface, the system isolates those deviations by constructing an **Observational Filtering Screen Matrix**.
When the data vector is passed through this matrix, all components of perfect information that match the original code of TLD are eliminated, forcing the **misaligned structures** to appear in their original form as a static data shadow, called **"Dark Matter,"** denoted by:
Based on the amplitude and structure of this "Dark Matter" block on the filtering screen, the system automatically decomposes the deviated information into two separate mathematical domains:

- **Region i) Mild Deviation (Linearly Correctable):** Located within the safe convergence radius. Here, the informational string is only slightly deviated from the canonical axis. The system can apply linear regression algorithms to pull the data back toward the perfect trajectory without damaging the underlying structure.
- **Region ii) Stubborn Deviation (Uncorrectable / Data Hell):** This is the dangerous region where the deviation exceeds the threshold and becomes more severe, particularly when the error vector falls into the **Null Space / Kernel** of the restoration matrix. **The kernel of a matrix** (also called the **null space**) is **the set of all input vectors that, when multiplied by that matrix, produce the zero vector**.

At this boundary, the determinant (a special real or complex number calculated from the elements of a **square matrix**, indicating the degree of change) of the structure collapses to absolute zero:
detrecovery=0
(det\det: determinant, an important value of a matrix.)
In matrix mathematics and theoretical physics, a zero determinant represents **permanent loss of information and creates non-invertibility**. Once it falls into this region, the system can no longer perform the inverse calculation needed to recover the original code; the informational sequence becomes completely isolated as a "data black hole." This is precisely the state of structural Hell, where information remains motionless in absolute

collapse with no way to save itself if it remains only within the t_1 plane.

b) The 4D Manifold t_2 and the Operating Mechanism of IGTM

To rescue information that is at risk of falling into the Null Space of t_1, the system is forced to activate a more complex evolutionary mechanism: **Information Pre-Geometrization and Temporalization into Matter (IGTM)** within 4D space t_1. This space does not develop from or borrow a pathway from t_1, but is generated directly from the origin t_0, choosing an evolutionary direction that is completely orthogonal and complementary to the Interface.

Whereas t_1 space is woven from straight, or slightly curved and twisted, linear strings, t_2 space is formed from **four special informational strings that curve and twist themselves but are arranged in an orderly manner**. Originating from the energy source t_0, these four strings extend outward and undergo the process of *Temporalization*—using an intrinsic time parameter as a driving force to coil around themselves in a helical spring-like structure. Each string is defined as a complex twisted vector field. The "orderliness" of these strings is ensured by their linear independence and structural orthogonality, keeping the 4D spatial framework stable and free from internal disorder.

The essence of the *Pre-Geometrization* process lies in the fact that **the twisting rate—the frequency ωj\omega_j of these informational strings—generates structural torsion (Topological Torsion Tμνλ, and this torsion is the very origin that generates mass and matter density.**

The mass generated from this twisted information follows the gravitational field through the geometric curvature of the surrounding 4D space according to the **Einstein Field Equation**. The ultimate purpose of this architecture is to **use matter to give form to errors**. When information carrying deviation enters 4D space, the twisting frequency of the strings changes, thereby creating a specific irregular deformation, a concrete gravitational anomaly in the physical world:

G=curvature=error and deviation.

$G_{\mu\nu}(\vec{\epsilon}) = G_{\mu\nu}^{\text{pure}} + \Delta G_{\mu\nu}(\vec{\epsilon})$)Experimental physicists may not be able to measure an abstract concept of informational deviation, but they can fully measure variations in matter and spatial curvature. By forcing an error to manifest as matter possessing weight and form, the system creates a dynamical environment in which intervening forces can gain purchase and carry out processes of reshaping and correction (as optimized through the Master Lagrangian function in Chapter 3).

In other words, **the error of information manifests as the weight of matter.**

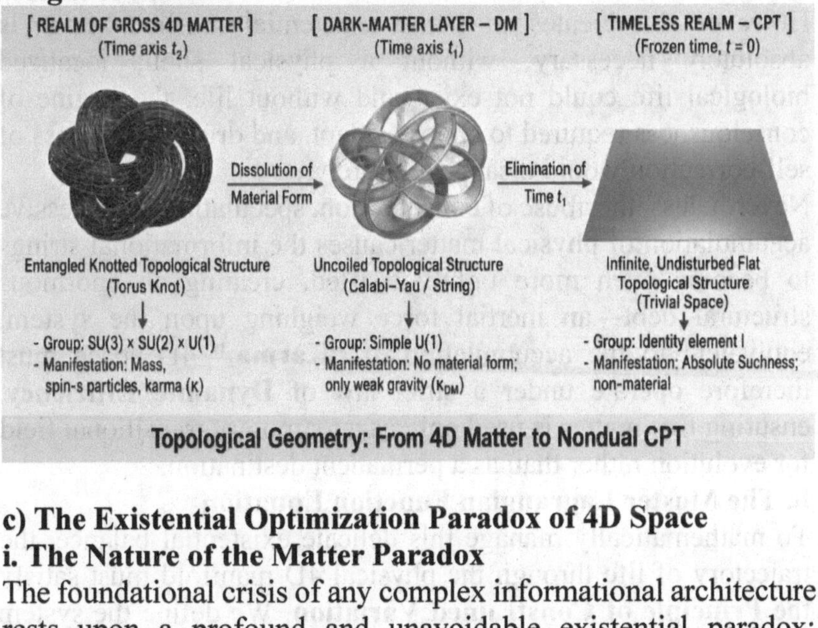

Topological Geometry: From 4D Matter to Nondual CPT

c) The Existential Optimization Paradox of 4D Space

i. The Nature of the Matter Paradox

The foundational crisis of any complex informational architecture rests upon a profound and unavoidable existential paradox: **abstract vector informational deviations** ε\varepsilon **cannot correct themselves in a state of absolute stillness**. In the flat Interface space (t1t_1), a structural error would be frozen; it would exist only as a static diagnostic trace—a shadow on a screen. If a deviation becomes sufficiently severe to approach the irreversible **Null Space / Kernel** region, it will face permanent systemic annihilation, because a flat and static environment possesses no dynamic "force levers" with which to manipulate, reshape, or redirect a collapsing data sequence.

To resolve this problem, the system deploys the process of **Information Pre-Geometrization and Temporalization into Matter (IGTM)**, pushing the erroneous informational state out of stasis and throwing it into the dynamic 4D manifold t_2. Here, the four fundamental informational strings of orthogonal 4D begin to twist with an intrinsic frequency (ω)(\omega). This intense twisting process generates structural torsion, physically manifested as mass density. $(G_{\mu\nu} = 8\pi G \cdot T_{\mu\nu}(\rho_{\text{matter}}))$

By wrapping an abstract error in heavy, tangible matter, the system forces an invisible defect to emerge into physical reality. However, this creates an intense existential conflict. Matter is absolutely necessary; without a physical shell, localized biological life could not exist, and without life, the engine of consciousness required to choose, adapt, and drive the process of self-correction would disappear entirely.

Nevertheless, the abuse of consumption, speculation, or excessive accumulation of physical matter causes the informational strings to become even more tightly twisted, creating an enormous structural debt—an inertial force weighing upon the system, equivalent to the accumulation of **"Karma."** 4D space must therefore operate under a strict law of **Dynamic Efficiency**, ensuring that matter is used only as a temporary transitional field for evolution rather than as a permanent destination.

ii. The Master Lagrangian Function Equation

To mathematically manage this delicate existential balance, the trajectory of life through the physical 4D manifold must satisfy the **Principle of Constrained Variation**. We define the system by a **Master Lagrangian function** based on IGTM theory.

According to **Hamilton's Principle of Least Action**, the true path of optimal evolutionary survival is the path along which the total variation of this universal action must equal absolute zero.

This variational optimization reveals a supreme universal law: **the physical 4D universe is mathematically designed to accept matter only in a temporary dynamic state**. Life is granted access to the physical picture in order to see its errors reflected in the form of matter, but it must immediately act to correct those deviations, unwind the local torsional frequencies $(\omega \to 0)$, and prepare the state to be flattened again into the pure Interface space t_1.

$$Benefit = (L(x, u, \tau) = J_{\text{heal}} - \lambda\, \text{karm}_{\text{a}})$$

iii. Spacetime Curvature as a Safety Net Against the Null Kernel Leading to Hell

This optimization framework explains why a descent into the physical world t_2, despite its serious karmic risks, is an act aimed at preserving the system. It serves as a **Thermodynamic Guardrail**, preventing life from crossing the event horizon and falling into the Null Kernel (Hell).

We embed this safety net directly into the material cost function by applying a nonlinear barrier method. When an entity slides deeply into profound error, the determinant of its fundamental structural restoration begins gradually decreasing toward zero. The material cost function is therefore rewritten to monitor proximity to this destruction:

Accumulated Karma

Karma is the antiderivative of the torsion of matter, the angle θ\theta, curvature, torsion, and the inverse determinant. When the determinant equals zero, the antiderivative collapses and karma becomes too great to be saved. This is the case when a living entity abuses its physical environment or refuses to correct its core informational deviations; it will drift toward the edge of the abyss where

$$K[\Gamma] = \int \Gamma \left[a_{\square_\kappa} (\kappa l) 2 + a_\tau (\tau l) 2 + / \, p \lambda B / |detJ|^p / \right] d\lambda$$

$\det(J_{\text{recovery}}) \to 0$

When

$\det(J_{\text{recovery}}) = 0$

karma $\to \infty$.

Hamilton's Principle prohibits any actual trajectory from reaching an infinite cost. This mathematical boundary therefore generates an extremely intense physical restorative pressure within the 4D world. On the material plane, it manifests as urgent crisis, inertial heaviness, suffering, or systemic collapse.

This physical burden is not a punishment; it is a vital diagnostic alarm. It forces the trajectory of life to pivot strongly, changing the control force u(τ)u in order to reverse the process of deterioration. The physical world operates as an enormous impulse absorber. By transforming an abstract, fatal data error into a dense and malleable material crisis, the 4D manifold intercepts a soul before it falls into permanent annihilation, granting it the time, weight, and evolutionary friction necessary to heal before returning to the **Timeless Domain**.

iv. Further Reading

When the pure informational flow moves along the angle θ\theta, curvature length κ\kappa, and torsion τ\tau, the matrix density is forced to acquire an additional **Berry geometric phase**, which rotates the matrix:

$(\rho(t) = U(t)\rho_0 U^\dagger (t)$

The density matrix (diagonal) dissolves comprehensive Knowing into individual fragments around the diagonal as the matrix rotates. This is the discrimination of the duality of **Manas consciousness.**

$V_{geometric}(\kappa)= -(\hbar^2/4m)\kappa^{-2}$ The curvature potential acts as an **attractive potential well**. The curvature κ\kappa of 4D spacetime creates the "folds" of ignorance. When the informational flow from TLD passes through these folds, it becomes localized in local **bound states**. As a consequence, information is no longer free to propagate but becomes condensed into fixed "data particles," forming the **Aggregate of Perception (Saṃjñā-skandha)** and **worldly knowledge.**

The greater the curvature κ\kappa (the heavier the ignorance), the deeper the potential well becomes, and the more information is distorted to conform to the form of its physical shell.

iv. Further Reading

When the pure informational flow moves along the angle θ\theta, curvature length κ\kappa, and torsion τ\tau, the matrix density is forced to acquire an additional **Berry geometric phase**, which rotates the matrix:

ρ(t)=U(t)ρ0U†(t).\rho(t)=U(t)\rho_0U^\dagger(t).

The density matrix (diagonal) dissolves comprehensive Knowing into individual fragments around the diagonal as the matrix rotates. This is the discrimination of the duality of **Manas consciousness**.

Vgeometric(κ)=−ℏ24mκ−2.V_{\mathrm{geometric}}(\kappa) = -\frac{\hbar^2}{4m}\kappa^{-2}.

The curvature potential acts as an **attractive potential well**. The curvature κ\kappa of 4D spacetime creates the "folds" of ignorance. When the informational flow from TLD passes through these folds, it becomes localized in local **bound states**. As a consequence, information is no longer free to propagate but becomes condensed into fixed "data particles," forming the **Aggregate of Perception (Saṃjñā-skandha)** and **worldly knowledge**.

The greater the curvature κ\kappa (the heavier the ignorance), the deeper the potential well becomes, and the more information is distorted to conform to the form of its physical shell.

Xin chuyen EN va khong thay doi ý: Toán tử tiến hóa biến ma trận mật độ thuần khiết thành ma trận mật độ

hỗn hợp (Mixed State Matrix):

The evolution operator transforms the pure density matrix into a **Mixed State Matrix**:

$\rho = \sum_i = \rho i \, |\Psi i\rangle\langle\Psi i|$ **Consequence:** Primordial Knowing is "diluted" and obscured by the probabilities of affliction Ψi\Psi_i. This is precisely the state of the **Aggregate of Consciousness (Vijñāna-skandha)** contaminated by defiled seeds.

Mathematical consequence: When mm becomes excessively large, the free kinetic-energy term is suppressed. The system loses its quantum flexibility (wave nature) and collapses completely into a classical particle (particle nature). Consciousness becomes condensed into the physical body, tightly confined within the potential well of Ignorance and subjected completely to the linear flow of time t_2 stratification requiring rebirth).

When the pure informational flow descends from TLD, **Manas consciousness (the Seventh Consciousness)** does not generate data. It performs only one function: **it clings to discrimination and appropriates the Eighth Consciousness (TLD) as the "Self."** In the state of Oneness (Plane Wave), Manas consciousness intervenes by imposing the inclination angle θ\theta. The matrix J(θ)J(\theta) immediately mixes the informational components, splitting reality into two opposing poles: **Self** (Subject) and **World** (Object).

The mass mm in this system represents **Karmic Density** or **Information Inertia**.

$$M(\theta, k, s) = \mathrm{Tr}\left(k_{\mu\nu} k^{\mu\nu}\right) + \alpha\, \mathrm{Tr}\left(s_{\mu\nu} s^{\mu\nu}\right)$$

The structural density of Ignorance is an **extrinsic-curvature tensor** or **polarization tensor** directly related to the curvature κ\kappa and the inclination angle, together with its conjugate tensor. Tr\operatorname{Tr} measures the **total amount of information being bent** when the flow from TLD is forced into finite 4D space.

This represents the degree of interaction with the mind's impulse of attachment —the ego's attachment to karmic consciousness.

- **When mm Is Large (Ordinary Being — Bound to the Physical 4D Realm)**

When the TLD structure accumulates too many defiled seeds (greed, anger, ignorance, attachment to views), the karmic density mm approaches a very large value and the kinetic energy is suppressed. The system loses its quantum flexibility (wave nature) and collapses completely into a classical particle (particle nature). Consciousness becomes condensed into the physical body, tightly confined within the potential well of Ignorance and subjected completely to the linear flow of time $t_2$stratification requiring rebirth).

- **The Process of Reducing the Density mm (Releasing the Seeds of Ignorance)**

Through the stages of concentration and wisdom, the practitioner "burns away" or filters the contaminated data particles within TLD. This is the process of **reducing the density of mental inertia**.

- **Transformation from Particle to Wave:** When attachments to knowledge and the reflexes of the Aggregate of Perception are eliminated, mm begins to decrease sharply toward 00.
- **Energy Release and Approach Toward Infinity:** Consider the geometric potential term:

$V_{geometric}(\kappa) = -(\hbar^2/4m)\kappa^{-2}$. As mm gradually decreases toward 00, if space still retains curvature κ\kappa (that is, if even a very small amount of ignorance remains, as in an Arhat or a Tenth-Bhūmi Bodhisattva), this potential well becomes extremely sensitive. This is why even a minute "inclination" remaining in the TLD realm of great saints still generates a very large corrective energy, requiring purification cycles tt to eliminate the error completely.

General Equation of Motion (When m→0m\rightarrow0)

When the informational flow remains confined by Ignorance and attachment to self, the equation of motion follows an extended Dirac form in curved space, where the karmic density

M(θ,κ,τ)M(\theta,\kappa,\tau)

plays the role of effective mass:

$$[i\hbar\gamma^s(\frac{\delta}{\delta x} - i\tau J(\theta)) - M(\theta,\kappa,\tau)c] \,|\, \Psi(\tau,t) = 0$$

Operating Mechanism of the Equation

When the function MM has a large value, the mass term MM becomes absolutely dominant. It acts as an **inertial resistance**, continuously "breaking" the wave function Ψ\Psi, forcing it to interact internally with itself. This process produces confined periodic oscillatory solutions—precisely the mechanism forming the repetitive cycles of the **Aggregate of Perception** and **worldly knowledge** (Knowing confined within the dualistic shell).

Coordinate Transformation Process (Reduction of the Probability of $t_1 t_2$Deviation)

When the practitioner practices meditation and enters deep states of concentration, the equation of motion undergoes a **Differential Transformation** intended to eliminate the deformation components:

1. **Reduction of psychological density:** The equation loses the perturbative component produced by the 4D environment.

2. **Purification of subtle t_1 habitual tendencies:** An Arhat or Tenth-Bhūmi Bodhisattva continues to cleanse geometric errors, bringing the structural tensor gradually toward its minimum level.

3. $[i\hbar\gamma^{s}]\frac{\delta}{\delta x}$ - -$M_{residual}$ c] | Ψ(τ,t)=0

The solutions of the equation at this stage are quantum waveforms with very broad spatial extension, transcending biological structure and expressing the higher-level Knowing of **"As It Is" (Wisdom).**

3. Ultimate State: The Massless Equation of Motion (m=0m=0)

When the practitioner attains Buddhahood, Ignorance and attachment to self are completely eliminated. The boundary conditions are established absolutely:

κ→0,τ→0,θ→0.\kappa\rightarrow0,\qquad \tau\rightarrow0,\qquad \theta\rightarrow0.

When karmic mass becomes zero, the equation of motion loses its mechanical structure and becomes:

$[i\hbar\gamma^{s}]\frac{\delta}{\delta x}$ - - | Ψ(τ,t)=0

Because the obstructing mass MM is completely absent, the equation has a true solution only when the wave function Ψ\Psi no longer depends upon the physical spatial coordinate τ\tau. 4D space is broken. The informational flow becomes an absolute constant with respect to space.

Transcending the time axis, t=0t=0: When spatial variation is no longer present, the time-evolution operator U(t)U(t) degenerates into the **Identity Operator**. The linear time axis completely disappears. The system enters directly into the **Timeless Domain (TLD).**

The Manifestation of Knowing "As Such"

The solution of the equation is now no longer a function of motion, but an **Invariant Pure State:**

F. CURVATURE OF SPACETIME: AN ONTOLOGICAL PERSPECTIVE

In Einstein's General Theory of Relativity, the field equation is written in the form:

$$(G_{\mu\nu} = 8\pi G \cdot T_{\mu\nu}(\rho_{\text{matter}}))$$

where G represents spatial distance, denoted by the coordinates μν\mu\nu, and T is the stress tensor.

This equation establishes the relationship between the geometric curvature tensor of spacetime $G_{\mu\nu}$ and the energy–momentum tensor $T_{\mu\nu}$. However, in essence, this is a mathematical relationship established from the principles of the theory, rather than an ontological explanation of the first cause of curvature. In other words, General Relativity describes **how the curvature of spacetime is related to the distribution of energy and momentum**, but does not aim to answer the question of **why spacetime itself has the capacity to curve**, or **what the deeper origin of that curvature is**. That is a question belonging to the domain of ontology, beyond the objective of the theory.

In the IPTM model proposed in this book, the curvature of spacetime is not regarded as a primordial reality, but as the consequence of a deeper structural layer: **informational curvature**. In the integral state of TLD, all informational structures are orthogonal, pure, and without curvature. When deviation from the primordial state of integrity appears, the informational structure develops curvature and the process of temporalization is activated. **This curvature occurs before matter or gravitational force appears.** The curvature of spacetime in the physical 4D world is therefore hypothesized to be the geometric manifestation of informational curvature at the foundational level.

From a philosophical and religious perspective, this deviation from the primordial state of integrity may be compared with the concept of **karma** in Buddhism or **original sin** in Christianity, because both imply a state that is no longer intact relative to the original foundation. However, this is only a philosophical comparison intended to help the reader visualize the concept, **not an identification of these concepts as being the same**. In Buddhism, karma is the consequence of intentional actions; in Christianity, original sin is a theological concept associated with the history of salvation. The IPPC model neither replaces nor reinterprets these doctrines, but merely proposes that both traditions may be reflecting, in different languages, a deeper principle concerning deviation from the primordial state of informational integrity.

Hell and the Black Hole: A Philosophical–Physical Correspondence

In the interpretation of the TLD model, *Hell* and the *black hole* may be placed side by side as two images of a state of *confinement*

—*confinement caused by curvature reaching an extremely high level*—but they are not the same reality. In physics, a black hole forms when matter–energy becomes sufficiently concentrated to create a region of spacetime with an event horizon; once matter or light has crossed that horizon, it cannot return to the outside along ordinary causal trajectories. In the TLD hypothesis, Hell may similarly be conceived as a *domain of informational confinement*, where the inclination θ\theta, informational curvature κ\kappa, or accumulated "corrective debt" has become too great for the structure to return easily to the state of integrity. Both therefore carry the image of *overload, extreme curvature, and difficulty escaping from the existing state*:

Black Hole : Hell (Total Degree of Loss):

Black Hole:

Black Hole:Hell (TLD):M/E↑⇒R spacetime ↑⇒Causal Confinement,θaccumulated↑⇒κinformation ↑⇒Corrective Confinement.

G. THE RETURN OPERATION AND DE-MATERIALIZATION

After a material entity in 4D space $t_2$has undergone a continuous process of reshaping and has successfully corrected its structural deviations through the control force u(τ), the system must perform the final step: returning the purified information to its original home, the Interface t_1. This transfer process requires a state transition from dynamic matter to static information without breaking or damaging the integrity of the data. This process is carried out through **Homotopy Flow** and a **Homeomorphism Transformation**.

When the deviated core has been healed, the system's curvature–torsion function automatically activates a continuous smoothing flow according to the curvature–torsion transfer parameter ss (running from 0 to 1):

- At the initial state s=0, the entity remains a curved and twisted material structure, subject to gravity in 4D.

- As ss gradually approaches 1, the system performs a process of **Adiabatic Spin-Down**. The informational strings begin to loosen their twists, lowering the coiling frequency toward zero:

loss of torsion→loss of matter.\text{loss of torsion} \rightarrow \text{loss of matter}.

From the perspective of theoretical physics, when the torsional frequency vanishes, the Cartan torsion and mass density of the entity completely disappear. Its Riemann curvature tensor becomes absolutely flat. This is precisely the process of **Dematerialization**: the entity sheds its heavy material shell, releases all gravitational forces, and returns to being a linear, weightless informational sequence.

At the point s=1s=1, the homotopy flow carries the entity through the gateway $t_0$and places it in complete alignment with the flat pathways of the Interface space t_1. The entity now satisfies the condition of absolute orthogonality with respect to the former 4D world.

All traces and memories of the deformation, suffering, or irregularities of the 4D material world have been completely erased through the projection matrix. The informational entity has been fully restored, reintegrated into the ordered flow of the Interface substrate, operating at a uniform velocity and remaining permanently immune to the twisting fluctuations of the surrounding physical manifold. The evolutionary cycle—from Origin, Deviation, Materialization and Error Correction, to the Return to Purity—has been closed in a mathematically perfect manner.

H. INTERFACE FIREWALL AND THE CONSERVATION OF ENTROPY

With **HTTM theory in the TLD model**, in order to ensure logic and consistency in the relationship between **TLD and the 4D world**, the exchange and transition between these two domains must necessarily be carried out through **defined intermediate mechanisms**. If HTTM theory is confirmed, these mechanisms would no longer carry only hypothetical significance, but would become necessary links for explaining the transformation from

pre-physical information in TLD into the **material and spacetime structures of the 4D world**, as well as the possibility of interaction in the reverse direction.

The mathematical equations proposed for the **Interface Domain** are therefore intended not only to reflect the internal consistency of the model, but also to describe a deeper natural mechanism: **the tendency to correct deviations in informational structures and return them to their original state of integrity and purity**. From this perspective, the mechanism for restoring order in the Interface Domain becomes the orienting foundation for processes occurring in the 4D world, creating conditions under which deviations that have arisen can be recognized, adjusted, and corrected.

However, this correction is not absolutely compulsory. Because **Freedom** is regarded as a primordial principle of the universe in the TLD model, human beings still possess the ability to choose and act according to their own intentions. It is precisely this capacity for choice that allows deviations to continue to arise, while at the same time opening the possibility of self-recognition and transformation. In human life, this tendency to return to a state of equilibrium is often manifested through **moral consciousness**, self-regulation, and the need to correct the consequences created by one's own actions.

To ensure that the original TLD server always maintains a static, eternal mathematical state with Entropy equal to 0, an ultimate filter is established at the dimensional boundary: the **Interface Firewall** is a natural but essential requirement.

a. Operator of the Purification Filter

When information moves through the 4D loop, it must pass through the filtering operator residing at the D6–D11 boundary.

TLD itself, in its potential state, contains all potential states simultaneously within an infinite-dimensional space. When passing through the Interface Domain to manifest as macroscopic reality, the Hermitian filtering operator acts as a **geometric projector** to extract information.

The operator is Hermitian,

$$(\rho_{\text{thựctại}} = F_F \cdot \rho_{\text{TLD}} \cdot F_F^{\dagger})$$

making the expression a perfectly symmetric filtering operation and ensuring that the density matrix after filtering remains normalized and physically real. (A Hermitian matrix is a matrix that returns to itself after transposition—rows become columns—and complex conjugation, i→−ii\rightarrow -i, and its determinant is a real number.)

The 4D universe is an open system that continuously generates disorder and thermal noise and increases entropy according to the Second Law of Thermodynamics. The Interface Firewall operates as a one-way filtering membrane:

i. It checks the symmetry and determinant of every ultimate information code (UIC) qubit sequence ascending from the synaptic matrix system.

ii. Only informational structures satisfying the condition of perfect topological symmetry are permitted to pass through Wick rotation and synchronize with TLD.

iii. If a component within the sequence loses phase coherence (**decoherence**) because of environmental noise, causing its determinant to approach zero, the matrix immediately performs an orthogonal projection to eliminate this erroneous element and returns it to physical space in the form of dissipated heat. (A matrix lists the properties of an event in a square or rectangular table arranged into rows and columns.)

b. UIC Information Density Through the Filtering Operator

The information density **IUICI_{\mathrm{UIC}} (Universal Information Concentration)** measures the amount of information within the Interface Domain. When it appears in the field equation, it is not a fixed constant; rather, it is precisely **the result of this filtering operation.**

f the definition of information in the form of **Von Neumann entropy** is used in combination with the filtering operator, we obtain the normalized expression for

$$I_{\mathrm{UIC}} = -\,\mathrm{Tr}\left[\left(F_F \rho_{\mathrm{TLD}}\right)\ln\left(F_F \rho_{\mathrm{TLD}}\right)\right])]$$

(Where Tr\operatorname{Tr} is the Trace operator—the trace of the matrix, taking the sum of the states after filtering.)

3. Direct Effect on the Extended Einstein Field Equation

When the filtering operator is substituted into the field equation at the Interface Region:

$$G_{MN} = 8\pi G \cdot T_{MN}\left(I_{\text{UIC}}\left[F_F\right]\right)$$

This mechanism explains how the physical world is shaped:

- It scans through the **Fibers (Fiber Space)** wrapped over the **TLD/TLD**foundation.
- It performs a sparse filtering operation, similar to the SPCA algorithm discussed previously, in order to select a set of **coherent states**.
- The distribution of these filtered states generates the informational density structure IUICI_{\mathrm{UIC}}.
- It is precisely this filtered informational density that generates the energy–stress tensor, forcing the higher-order manifold structure to expand and contract, thereby producing geometric curvature—gravity and induced spacetime.
- Because the filtering operator is a **Hermitian operator** acting directly on the fiber structure of TLD, it **does not depend on the time parameter tt**. (Hermitian means a matrix that always gives real numbers; an operator is a mathematical object that acts to control the equation.)
- This filter does not "run" with time; it exists as a fixed geometric structure at the Interface Region. The nature of the "flow of time" that we perceive is only the local variation of the eigenvalues as we move the geometric connection along different representation domains of the filtering operator.

I. THE REALITY-FILTERING MECHANISM: How the Universe Eliminates "Deviated Illusions" to Return to the Eternal Ontological Reality

Imagine **Ultimate Knowing (the Timeless Domain — TLD)** as a still, transparent ocean containing all the films of the universe. In this ocean, time does not flow; the past, present, and future of all things exist simultaneously and completely in the form of an ultimate source code called **UIC**. However, when this code passes through the Interface Region to manifest as the material world in which we live, it must pass through an invisible but absolutely rigorous geometric filter: the **Hermitian Filtering Operator**.

During this process of extracting information, the flow of cognition is disturbed, generating noise impulses and forms of "deviated information." These are mistaken perspectives, fears, frequency disturbances, or temporary illusions generated by the localized self. In mathematical–physical terms, these deviated forms of information carry the values of "purely imaginary numbers"—they are not real at the core ontological level, yet they create pressure that distorts spacetime.

At this point, the **Hermitian Filtering Operator** operates like a perfect light-polarizing prism. Because of its absolute self-symmetric property, this filter permits only what belongs to "Actual Reality" (real numbers) to pass through and construct stable matter and energy. Conversely, all deviated information and asymmetric illusions are blocked by the filter directly at the gateway of the Interface Region. Because they are fundamentally groundless and out of phase with the UIC source code, they permanently fall into the **Null Space**—a mathematical region in which they are completely annihilated to zero.

It is precisely this ruthless and perfect filtering mechanism that ensures that everything damaged, deviated, and temporary in the macroscopic world is crushed, erased without trace, and **can never return to or contaminate the Timeless Domain (TLD)**. The ocean of Knowing thereby eternally maintains its purity, stillness, and invariance throughout countless lifetimes.

a) Separation of the "Real" and the "Deviated" Components

Within the Hilbert space of the fibers (**Fiber Space**), any raw informational operator or data matrix containing both canonical information and deviated information can be uniquely decomposed into two components: $M = M_{\text{Hermitian}} = M_{\text{Skew-Hermitian}}$

Where:

- The **Hermitian part** is the perfectly symmetric component, representing objective and invariant information of reality.
- The **deviated part (Anti-Hermitian / Skew-Hermitian)** represents phase-error noise, illusory components, or

asymmetric variations that do not belong to the eternal ontological nature of the Timeless Domain.

b) Mechanism for Filtering Deviated Information

When the filtering operator acts upon the system, it uses **geometric orthogonality** to eliminate the deviated component through three fundamental steps.

Forcing the Eigenvalues onto the Real Axis

A deviated operator intrinsically possesses **purely imaginary eigenvalues**.

The algebraic structure of the Hermitian filter accepts only **real eigenvalues**.

When a quantum measurement or data-filtering operation is performed, the deviated information—the purely imaginary component—falls into the **Kernel/Null Space** of the filter. In other words, its probability of manifestation becomes zero, and it is completely eliminated at the Interface Region.

Orthogonal Projection

If the filter is structured as a **Hermitian Projection Operator**, it divides the fiber space into two completely orthogonal subspaces:

- **Preserved subspace:** contains the canonical UIC source code.
- **Deviated subspace:** contains geometric deviations.

The filtering process is performed by projecting the raw state matrix through the filter, leaving the resulting state completely free of deviated components.

The **F_F filter** in meditation operates exactly like the **filtering membrane of an artificial kidney** or the **specific antibodies of the immune system**. The raw density matrix ρraw\rho_{\mathrm{raw}} is the bloodstream filled with toxins, impurities, and inflammatory agents—the skew components entering from the external world. When this blood passes through the FF membrane, the quantum antibodies identify and immediately neutralize these toxins, reducing them to zero:

$$(\rho_{\text{local}} = F_F \cdot \rho_{\text{skew}} \cdot F_F$$

The blood emerging from the filter becomes completely purified and free of toxins, protecting the life and primordial purity of the entire mind–consciousness system.

c) Elimination of Off-Diagonal Elements (Decoherence of Deviated Noise)

Deviated information is usually located in the **off-diagonal terms** of the informational density matrix, representing phase differences between the fibers.

Because the filtering operator commutes with the global Hamiltonian, it forces the system to maintain its state in a static **eigenbasis**. This process is analogous to **quantum decoherence**, but occurs geometrically: the filter removes random deviated phases and retains only the pure informational density.

The **Mindfulness Filtering Operator F_F** must be a self-adjoint Hermitian operator:

$F†F= F_F$.

This property ensures that the eigenvalues of the filter are always real numbers,

$\lambda \in R$,

representing empirically realizable states that can be directly measured through mindful experience.

Physical and Ontological Meaning of the Hermitian Property $F†_F= F_F$

In quantum mathematics, the operation consists of reversing the matrix—**transposition**, converting rows into columns—and changing the sign of the imaginary part through **complex conjugation**. When an operator satisfies

$F†F=F$,

it carries two important meanings.

Ensuring that the result is a "Real Number" (No Illusion)

If an operator is not Hermitian, its eigenvalues will be complex numbers containing the imaginary unit ii; ii is the driving element that produces rotation, oscillatory waves, and illusion.

- **If the filter contains ii:** the filter itself becomes distorted, rotates and twists itself, and generates illusion. The filtering operation becomes biased according to the observer's mood or preconceptions.

- **When $F^\dagger_F = F_F$:** mathematics proves that 100% of its eigenvalues must be **Real Numbers**. In meditation, this means that the filter is completely objective and realistic, "seeing things as they are" (**Vipassana**), without being mixed with any complex illusion of the dualistic mind.

Ensuring Conservation and Justice (Law of Cause and Effect)

A Hermitian operator preserves the global probability structure. When the matrix is projected through FFFF at both ends, the Hermitian property ensures that the filter removes only the deviated information—the skew component—while **the canonical information, the pure Knowing within, is neither lost nor distorted**. It operates like an absolutely precise scale of justice.

"A filter satisfying the Hermitian property $F^\dagger_F = F_F$ *is like a biological membrane with 100% perfect compatibility with the body, or a laboratory-grade collimating spectral filter. This filter does not add its own color to the object and does not distort the image; it is completely transparent and neutral. Precisely because it is self-adjoint (Hermitian), it creates neither friction nor side effects within the system. Its sole function is to allow the direct light of the Nature of Suchness to pass through while retaining and eliminating all the deviated rays of Ignorance."*

Physical Meaning for the Extended Field Equation

Without the filtering operator to eliminate deviation:

- The informational density would contain complex/imaginary components.
- The energy–information tensor would consequently contain these components.
- This would violate the geometric property of the Einstein tensor GμνG_{\mu\nu}, which is always symmetric, causing the spacetime structure at the Interface Region to collapse or generating physical paradoxes such as imaginary energy flows or unstable wormholes.

Because the filtering operator cleanses the deviated information, the informational density always remains a normalized real-valued density function, ensuring that the right-hand side remains perfectly symmetric and allowing induced spacetime to emerge stably while obeying global conservation laws.

ASYMMETRY BETWEEN CONTROL FROM TLD AND FEEDBACK FROM 4D

Although the 4D world may be imagined as a *performance stage* where information from TLD is actualized, **the flow of information between TLD and the 4D world is not symmetric.**

According to the hypothesis of the model, in the direction from **TLD to 4D**, the structural information necessary for actualization is transmitted fully through the temporalization process. It acts as the foundational code that determines the structure and conditions for the manifestation of physical reality.

Conversely, in the direction from **4D back to TLD**, feedback is not transmitted back in its entirety. Experiences, transformations, and data generated within the 4D world must first pass through the **Interface Firewall**. This mechanism acts as a filter, allowing only those informational components that satisfy certain conditions to return to the deeper domain, while noisy, inappropriate, or non-qualifying information is restricted or eliminated. Thus, the exchange of information may be represented as:

$$\mathrm{TLD} \xrightarrow{t\ encoding} \mathrm{4D} \xrightarrow{4\,DFeedback} \mathrm{FF} \xrightarrow{Selective\ Filtering} \mathrm{TLD}$$

in the reverse direction, where FFFF is the **Interface Firewall.**

Another example is that AI can now transform static information into a video. A still image can be transformed by AI into a video if appropriate software is used. The software represents the process of temporalization.

According to the TLDmodel, the relationship between the **proto-individual** and the 4D individual is similar. The individual in the phenomenal world is the manifestation of a deeper informational structure. The existence and development of the individual are always connected with the proto-individual, while the ability to directly perceive the proto-individual is limited by the very conditions of spacetime.

Therefore, the path of cultivation, development of wisdom, and purification of information does not aim to create a new entity, but rather to help the phenomenal individual become increasingly aligned with the integral structure of the prototype in TLD. In the theatrical-script analogy, the actor can also provide feedback to the director and scriptwriter when necessary. This is the process of self-correction. As informational deviation is gradually transformed, the distance between the phenomenal individual and the proto-individual is gradually reduced.

DECODING PHENOMENA AT THE BOUNDARY OF REALITY

Sixth Sense and Information Leakage Through the Interface Domain

As presented above, within the proposed model, **TLD**, from a philosophical and ontological perspective, is regarded as the deep foundation of the phenomena manifested in the 4D world. From this perspective, every phenomenon in 4D may be regarded as a manifested image of an informational structure lying deeper within TLD.

This may be visualized through an analogy: just as a playwright can follow the entire progression of the play that they have constructed, TLD continuously maintains an informational relationship with the processes manifested in 4D. However, the reverse relationship does not possess equivalent symmetry: entities in 4D ordinarily cannot directly access all of the information belonging to the foundational domain that generated them.

According to the TLD–4D hypothesis, the connection from TLD to 4D is foundational because the informational structures of TLD are assumed to participate in the very formation and operation of 4D phenomena. This connection may extend from microscopic to macroscopic scales.

At the quantum level, phenomena such as **quantum entanglement** and the **double-slit experiment** show that the microscopic world cannot always be described as a collection of classical objects possessing properties completely independent of the measurement process. In the double-slit experiment, for example, when information about the particle's path becomes available through an appropriate measurement interaction, the interference pattern may be reduced or disappear. This means that the photon "knows" that a human is observing it; or more precisely, the quantum state depends on the interaction of time acting as a physical factor, transforming information into a particle along its path and under the conditions of measurement.

On the basis of the hypothesized **TLD–Interface Domain–4D** relationship, the model proposes that information about the state of a system in 4D may continuously be correlated with its corresponding informational structure in TLD through the Interface Domain. TLD may be compared to a foundational "information server": the manifested state in 4D is updated into the foundational structure, while structural conditions from TLD may act back upon the process through which the system manifests in 4D. This is only an analogy intended to express the hypothesis; it does not imply that TLD is literally a computer or physical server in the ordinary sense.

However, the exchange of information between the two domains is hypothesized to be **asymmetric**. The direction from TLD to 4D is regarded as strong and continuous, whereas the direction allowing a subject in 4D to directly access information from TLD is highly restricted.

Nevertheless, this asymmetry does not necessarily have to be absolute. The model hypothesizes that, under special conditions, a certain amount of information that is not directly available to ordinary perception in 4D may cross the Interface Domain and become available to a 4D subject.

If such a mechanism truly exists, the TLD model calls this phenomenon **Interface Information Leakage**. At the level of subjective human experience, certain phenomena commonly called the **sixth sense, unusual intuition, or extrasensory perception** may be proposed as candidates for investigation under this hypothesis.

However, this interpretation currently belongs only to the scope of the **TLD model hypothesis** and is not a conclusion confirmed by experimental physics. To move from an ontological model to a scientifically testable theory, it would be necessary to identify the mechanism of information transmission, its limits, and experimental predictions capable of distinguishing it from chance, subconscious inference, and ordinary cognitive mechanisms.

9. QUANTUM DISINTEGRATION PROCESS OF THE FIVE AGGREGATES AT DEATH: AN EXPLANATION FOR THE 21-GRAM EXPERIMENT

The phenomenon of abnormal weight loss and fluctuations at the moment of death in Duncan MacDougall's twentieth-century experiment is scientifically decoded through the **Progressive Quantum Phase Transition** of the structure of the Five Aggregates.

The Soul or Knowing is not a single homogeneous mass of matter, but a tightly interconnected stratified system extending from D6 to D11. When the process of biological death is completed, disconnection does not occur instantaneously throughout the entire system, but collapses through independent phases:

a) Phase 1: Dissolution of D6 (Form Aggregate): The energetic scaffold composed of the Dark Substrate and the flow of Neutrino particles becomes completely disconnected from the peripheral nerve endings and bodily synapses. Billions of free Neutrino particles escape from the gravitational force of D5 and return to a state of free informational waves, producing the first empirically observable decrease in mass.

b) Phase 2: Collapse of D7 & D8 (Feeling & Perception Aggregates): The raw sensory matrix and the automatic reflex storage system are disconnected from their energy source. The dissolution of these microscopic geometric connections produces temporary fluctuations, increases, and decreases in weight due to the sudden release of quantum entropy impulses at the Interface boundary, indirectly affecting the background gravitational force of D5. This perfectly explains why the measured weight of corpses fluctuated chaotically in MacDougall's laboratory.

c) Phase 3: Retrieval of D9 & D10 (Formation & Consciousness Aggregates): The final UIC qubit-code sequence storing the complete memory matrix, behavioral imprints, and the consciousness operating system is packaged and compressed into the most condensed topological structure.

d) Phase 4: Transcendence into the Eternal Domain: Knowing (D11) receives the purified data core of Consciousness and performs a permanent Wick Rotation. The real-time coordinate axis tt closes, carrying this structural informational entity completely through the Interface Firewall and permanently retrieving it into the static, invariant, and omniscient mathematical state of the original TLD Server.

According to HTTV/IPTM, death is not the termination of consciousness, but a **great data-retrieval algorithm**, returning the quantum software, after one operational cycle in the material 4D world, to complete synchronization with the primordial source of **Primordial Wisdom** in the Timeless Domain.

STRATIFICATION DIAGRAM FROM TLD TO THE 4D WORLD

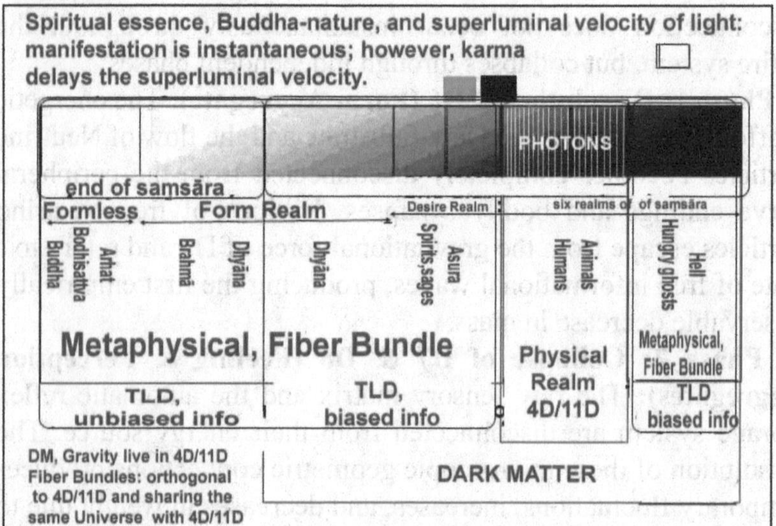

The **black region** represents **dark matter**, with the thin time t_1, belonging to the **Interface Domain**, where **karma** exists and operates.

The **white-striped region** represents **photons and the 4D material world**, with full physical time t_2.

Hell and the Realm of Hungry Ghosts consist only of dark matter and contain no photons. In these realms, information possesses **maximal curvature and torsion**, hypothesized to approach physical conditions analogous to those of a **black hole**. Above these regions, the **Human Realm and Heavenly Realms** still remain within the cycle of samsara and continue to be governed by karma, although some of these domains may contain no photons.

The **Formless Realm and the upper portion of the Form Realm** are represented as domains of pure information, where informational deviation is absent or nearly absent. The light in these realms is not necessarily photon-based light; rather, the model hypothesizes it as the **primordial light of Wisdom**.

CHAPTER IV: THE RELATIONSHIP BETWEEN HTTV/IPTM THEORY AND STRING THEORY

Summary

The Relationship between HTTV Theory (Pre-Geometrization and Temporalization of Information into Matter – IPTM: Information Pre-geometrization and Temporalization into Matter) and String Theory

String Theory is one of the most important attempts to unify quantum mechanics and general relativity. Although it has achieved many mathematical advances and introduced new approaches to several paradoxes of modern physics, it is still not regarded as a Theory of Everything. Many fundamental questions remain open, such as: Why does spacetime exist? Why can strings vibrate? What is the origin of information? What determines the state of the observer in quantum phenomena?

The KPT/TLD model proposed in this book hypothesizes that behind spacetime and quantum fields there exists a timeless informational domain (TLD), which serves as the foundation of all phenomena. On this basis, the model proposes a new approach to explaining certain quantum paradoxes, while also suggesting that the nature of strings and their vibrational modes may be related to three fundamental geometric parameters of information: inclination θ, curvature κ, and torsion τ. These parameters are regarded as quantities describing the degree of deviation from the state of informational integrity within the TLD, thereby leading to the process of temporalization and actualization into matter.

In other words, information in TLD is the pre-geometric precursor of inclination, curvature, and torsion, corresponding to different vibrations and directions of string vibration:

$\omega=\Omega(\theta,\kappa,\tau)$

(vibration), corresponding to the energy bridge according to the function.

In addition, the model also proposes that the Interface Domain between TLD and 4D spacetime may provide a theoretical framework for interpreting the role of the extra dimensions in String Theory and M-theory. According to this hypothesis, the additional dimensions from D5 to D11 are not merely mathematical dimensions required to ensure the consistency of the theory, but may also correspond to different levels of informational transformation in the process of material actualization.

In the proposed interpretation, dimension D5 may be associated with the minimal-time domain t_1*, where gravity and dark matter begin to emerge; the subsequent dimensions may be correlated with the process of manifestation of the Five Aggregates in the following order: D6 corresponds to Form, D7 to Feeling, D8 to Perception, D9 to Mental Formations, and D10 to Consciousness. This correspondence is not intended to assert an identity between Buddhism and modern physics, but is proposed only as an interdisciplinary model seeking a common language for describing the process of transformation from primordial information to the phenomenal world.*

If this approach has a valid basis, the HTTV/IPTM model would not replace String Theory, but could serve as a pre-geometric foundational layer, contributing to an explanation of the origin of spacetime, vibrational modes, and the role of information in the formation of matter and consciousness.

Although **String Theory remains the subject of considerable debate** and has not been accepted by the entire theoretical physics community, it is still regarded as one of the most ambitious attempts to unify the four fundamental interactions of nature within a single mathematical framework. For this reason, String Theory has become an important foundation for many approaches toward a "Theory of Everything."

However, alongside its mathematical successes, String Theory still faces many fundamental questions. For example, the theory has not provided a clear mechanism to explain the role of the observer in quantum mechanics, nor has it provided a widely accepted interpretation of phenomena such as quantum superposition, quantum tunneling, or wave-function collapse. Beyond these, ontological questions such as the origin of spacetime, the origin of information, or the underlying cause that makes strings vibrate still remain outside the explanatory scope of the theory.

HTTV THEORY (PRE-GEOMETRIZATION AND TEMPORALIZATION OF INFORMATION INTO MATTER) / IPTM

According to the perspective proposed in this book, this difficulty does not necessarily arise from String Theory itself, but may reflect the fact that contemporary physics and mathematics are still missing a deeper foundational layer. When physical phenomena are regarded as ultimate reality, questions concerning the nature of existence inevitably emerge as paradoxes or unanswered questions, such as: Why does spacetime exist? Why do strings vibrate? Where does information come from? What determines the state of the observer?

In the TLD model, these questions are approached from a different ontological foundation. TLD is hypothesized to be a primordial, timeless, and pre-geometric informational domain in which matter, fields, and spacetime do not yet exist. On this basis, IPTM (**Information Pre-geometrization and Temporalization into Matter**) is proposed as a mechanism describing the process by which primordial information is progressively geometrized, stratified, and actualized into the material world.

From this perspective, IPTM does not replace String Theory but serves as a deeper theoretical layer describing the stage before strings and geometric structures emerge. If String Theory explains how string vibrations generate the spectrum of elementary particles, IPTM attempts to answer two questions that remain open: **what creates the strings, and what causes the strings to begin vibrating?** In other words, IPTM serves as a bridge between the ontological foundation of TLD and the geometric descriptions of String Theory, thereby forming a continuous chain of reasoning:

TLD/TLD→ IPTM→Vibrating Strings

$\theta \rightarrow \kappa \rightarrow \tau \rightarrow \omega \rightarrow$particle

In other words, according to the TLD model proposed in this book, informational structures in the timeless domain may be regarded as the **pre-geometry** of physical objects in 4D spacetime. Before matter, fields, and spacetime appear, information already exists in the form of primordial geometric structures characterized by inclination θ\theta, curvature κ\kappa, and torsion τ\tau.

When the process of temporalization is activated, these pre-geometric structures are actualized into physical objects in the 4D world. According to the hypothesis of the model, these three parameters may determine the geometric and dynamical characteristics of strings, such as curvature radius, direction of vibration, and vibrational modes. Therefore, the vibrational frequency of a string is no longer an independent parameter, but may be regarded as a function of the informational structure:

$\omega=\Omega(\theta,\kappa,\tau)$(Vibration).

Thus, TLD proposes a deeper order in which the matrix operator, the Hamiltonian H, describes the dynamics of the system. According to the interpretation of the TLD model, the Hamiltonian not only reflects the energy of the string in spacetime, but may also be regarded as a manifestation of the pre-geometric informational structure after it has been temporalized. Thus, the vibrational modes of strings in the 4D world may be dynamical projections of deeper informational structures within the TLD.

1. From Information to Ethics

A distinctive feature of TLD is that it goes deeper than String Theory in that it **does not separate physics from ethics, or matter from spirit**.

In standard physics, deviation, curvature, spin, and vibrational modes are merely mathematical quantities. *But in TLD, they may carry a deeper meaning related to ethical principles. Deviated information is a separation from ethical standards such as the Ten Commandments in Christianity or the precepts in Buddhism.*

Integral information is information that has not deviated from its primordial nature. When deviation occurs, curvature appears. When curvature appears, gravity, mass, matter, and constraint emerge.

At the spiritual level, this deviation manifests as greed, anger, ignorance, pride, doubt, and attachment.

At the physical level, it manifests as mass, forces, fields, and matter.

Therefore, an integrated chain may be written as:

Pure Structural Information→Δθ→κ,τ→ω→Field→Particle→Matter→Materialistic Attachment

From this perspective, vibration is not the first cause. Vibration is the consequence of a deeper informational geometry.

2. When **Wick rotation** is activated at the Interface boundary, the static informational distribution function of TLD is immediately pushed into a quantum dynamical phase.

Informational deviation makes the system dynamic, represented by HH (the Hamiltonian), which is the operator of the total energy of the system and functions to indicate the energy/degree of manifestation of the quantum system upon which it acts.

$e^{-\frac{iHt}{\hbar}}$ transforms imaginary time τ into t, bringing the system to life.

The equation above is precisely the bridge that makes the strings in String Theory move and generate elementary particles.

This transformation converts the static phase constant θ\theta in TLD into an angular frequency varying with real time, while simultaneously releasing the wave vector κ\kappa into physical momentum. Real time is born not from a mechanical explosion of matter, but from the very moment when the complex coordinate axis of TLD "rotates" through Wick rotation.

With this chain of reasoning, TLD, IPTM, and String Theory are not competing theories, but three different descriptive layers of the same process of actualization: **TLD** answers the ontological question, **IPTM** describes the mechanism of transformation from information into geometry, while **String Theory** describes the dynamics of the geometric structures that have already formed.

Returning to the Beginning of String Theory: Redefining String Theory — The Energy String Is Data Pressure from TLD

String Theory and its extension, M-Theory, demonstrate that the elementary particles of matter are not point particles, but vibrating loops of energy in 11-dimensional space (11D). Nevertheless, String Theory remains at an impasse because it cannot answer what the string itself is made of and why it naturally vibrates.

The hypothesis of **IPTM (Information Pregeometrization and Temporalization into Matter)** emerges to provide this fundamental ontological answer. The string is not a self-generated material entity. The essence of the string is **data pressure from TLD flowing across the Interface Region**.

When the informational flow containing the three codes θ,κ,τ passes through Wick rotation, it generates enormous mathematical pressure, forcing spacetime at the boundary to expand, contract, and curve. The extended Einstein field equation at the Interface Region is established on the basis of the informational density of the UIC code (**Universal Information Code**).

Here, topological curvature is forcibly shaped into microscopic folds at the Planck scale. To an observer in physical space, these topological folds woven from informational pressure manifest precisely as **strings**. The vibrational modes of the strings are in fact the dynamical projections of the three precursor codes θ,κ,τ as they are decoded through the geometric structure of space.

3. M-Theory as the Interface of TLD

In this interpretation, M-Theory is no longer the ultimate foundational layer.

It is the first Interface layer that can be described mathematically.

MM may be understood as an informational field located at the boundary between TLD and the manifested world. Branes are not necessarily matter.

They are multidimensional informational fields. Strings are one-dimensional informational structures existing within that field. Therefore:

String corresponds to dynamic information.

Brane corresponds to the informational field.

M corresponds to the Interface Domain of TLD.

In this sense, Yang–Mills theory, quantum fields, and even spacetime itself may all be secondary manifestations of a deeper informational layer.

4. Separation of the 11 Dimensions: The True Role of the Hidden Spatial Dimensions

M-Theory requires the universe to have exactly 11 dimensions so that the mathematics remains free of inconsistency. Of these, four dimensions—three spatial dimensions and one time dimension—expand into the macroscopic universe, while six additional spatial dimensions, from the sixth to the eleventh dimension, are tightly compactified into microscopic Calabi–Yau geometries.

Traditional physics regards these six hidden dimensions as concealed physical spaces. IPPT reverses this perspective and asserts: **these six hidden dimensions are not physical space, but are the informational coordinate axes of the Interface Region.**

The mechanical asymmetry between the two groups of dimensions demonstrates this:

- **Manifested dimensions (D1–D5):** This is where gravity, residing at D5, and the other forces of nature operate, allowing the vibrational modes of strings to transform into particles and dynamics that humans can measure and experimentally verify in 4D physics.
- **Informational dimensions (D6–D11):** Here, no additional mechanical force of nature exists. These dimensions are completely devoid of material dynamics and appear only through purely mathematical equations. This is because these six dimensions function as a **stratified microprocessor** that decodes each aggregate of life: D6 (Form), D7 (Feeling), D8 (Perception), D9 (Mental Formations), D10 (Consciousness), and D11 (Knowing/Soul). These hidden dimensions contain the algorithmic control structure of the entire universe and living world.

5. The Nature of Strings: Where Do They Come From?

String Theory: It regards elementary particles not as point particles but as vibrating loops of energy. However, the theory **does not explain** what the string itself is fundamentally made of. It assumes the string to be the most fundamental entity.

Similarity with GMGT: If GMGT is placed at the foundation, **structural quantum information in TLD provides the answer**. The informational modes θ,κ,τ are not matter, but when they interact and oscillate, they "manifest" in the physical world as geometric entities. The strings of String Theory are precisely geometric structures woven from this TLD information.

6. Precursors of the Modes of Vibration

String Theory: The frequency and configuration of string vibrations determine the mass, charge, and spin of particles—for example, one vibrational mode produces a photon, while another produces a graviton.

In GMGT: Geometric/informational parameters such as the phase angle θ, curvature κ, and entropy/spin τ serve as **precursor codes**. When these informational codes operate within the TLD pre-space, they force spacetime to curve or oscillate according to specific mathematical ratios. Therefore, the vibrational modes of strings are actually surface manifestations of the deeper mathematical modes θ,κ,τ\theta,\kappa,\tau originating from TLD.

7. Difference in Space: The Entire TLD Domain versus 11 Dimensions (11D)

String Theory (M-Theory): Requires 11 dimensions—or 10 dimensions in superstring theory—for mathematical consistency and the cancellation of anomalies. These extra dimensions must be tightly compactified into complex geometries such as Calabi–Yau spaces. Strings still vibrate *within* this multidimensional spatial framework.

GMGT Model: Information resides throughout the **entire TLD (Timeless Domain)**. TLD is essentially an extension of real space by imaginary coordinates, which are the origin of the informational field and quantum phase.

Point of Connection: The 11-dimensional geometric structures or compactified Calabi–Yau spaces of String Theory may themselves be **projections**, or the way mathematical physics attempts to "physicalize," the complex informational dimensions of TLD into real space. Instead of requiring 11 dimensions of physical space, we need only TLD space with self-consistent geometric informational structures.

MECHANISM FOR TRANSFORMING INFORMATION INTO ELEMENTARY MATERIAL PARTICLES WITHIN THE INTERFACE AND TIMELESS DOMAINS

When time begins to emerge with t_1, deviated and twisted structural information is pushed down into the Interface Domain approaching timeless 4D space.

Within the Interface region between **Quantum Field Theory (QFT)** and **11-dimensional space (11D M-Theory)**, the "code" that excites the vibration of strings is no longer an ordinary mechanical force, but a purely geometric and informational forcing. The nature of reality at this boundary may be interpreted through an integration of theoretical physics and quantum philosophy.

At this interdimensional boundary, spacetime is no longer a static stage but becomes a dynamical entity in which **quantum information is inclined, curved, and twisted**. The twisting of the time dimension and the deformation of configurational geometry generate a profound parallel-transport effect. As time-evolution operators transform the density matrix along a nonlinear trajectory, the quantum system accumulates a macroscopic **Berry geometric phase** *(Berry phase: a change in the state of an entity within a larger system when that entity moves along a closed path in order to return to its initial state)*. This geometric phase itself acts as a forced effective potential, breaking the equilibrium of the underlying vacuum state.

From a philosophical perspective, 11D space acts as a "primordial source" containing all informational codes in a state of ultimate potentiality, while the TLD Interface region is where this informational structure is compressed, decoded, and refracted. When the informational flow crossing this boundary is bent by the **Metric Tensor** structure, the vacuum fluctuation energy undergoes local symmetry breaking. This asymmetry acts as an ultimate "crank mechanism," striking the fundamental strings stretched across the interdimensional Interface. The strings are forced to change their vibrational frequencies, transforming geometric impulses and deviated informational configurations into physical notes.

HYPOTHETICAL SPACE BEYOND THE LIGHT CONE: A PROPOSED CORRESPONDENCE

In the TLD model, a distinction must be made among the **light cone (LC), dark matter (DM), and the Hubble sphere**. The light cone belongs to the causal structure of four-dimensional spacetime and is defined by null trajectories, $ds^2=0$ with the local speed of light (c). Photons propagate along null trajectories that define the boundary of the light cone.

Within the TLD hypothesis, **baryonic matter (BM) and photons belong to the full-time level (t_2)**, whereas dark matter is associated with the minimal-time level (t_1), or **"thin time,"** and does not require photons as a constituent or as a mechanism of materialization. In this conceptual sense, the light-cone domain is associated with the photon-accessible (t_2) physical world containing baryonic matter, whereas DM is associated with the (t_1) sector of 4D manifestation, characterized in TLD by thin time and the absence of photons as a necessary component of its materialization.

The **light cone must also be distinguished from the Hubble sphere**. The Hubble sphere is defined by the Hubble radius

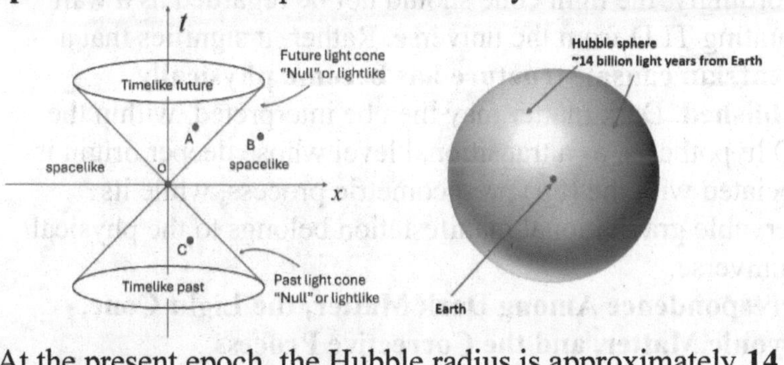

At the present epoch, the Hubble radius is approximately **14 billion light-years**. Galaxies beyond the Hubble sphere can therefore have recession velocities greater than (c). This does not mean that they lie outside photon-defined or observable spacetime. Rather, their superluminal recession results from the expansion of cosmic proper distance. The Hubble sphere is therefore neither the light cone nor the boundary of the observable universe.

From TLD to the 4D Causal Structure

Within the TLD model, one may hypothesize a transition from a **pre-geometric informational domain** toward the physical 4D world and its causal structure.

TLD should not be directly identified with the spacelike region of relativity, because TLD is postulated to be **timeless and pre-geometric**, where neither a Lorentzian metric nor a physical light cone has yet been established. Nevertheless, the region conventionally represented as **outside the light cone** may serve as an intermediate mathematical analogy suggesting the absence of ordinary 4D causal connectivity.

When temporalization occurs at the minimal level (t_1), information from TLD is hypothesized to enter a **pre-material dark sector** and subsequently manifest as dark matter. Once dark matter becomes physically manifested, however, it should no longer be regarded as spacelike matter. Its observable gravitational manifestation must participate in 4D spacetime and exhibit the appropriate causal behavior.

$$\nearrow \quad t_1 \rightarrow DM$$
$$TLD — t_0 \quad \angle_{\rightarrow > \square}\, proto\ \ t2 \rightarrow protoBM(beyondLC) \rightarrow BM(LC)$$

Accordingly, the light cone should not be regarded as a wall separating TLD from the universe. Rather, it signifies that a **Lorentzian causal structure has become physically established**. Dark matter may then be interpreted, within the TLD hypothesis, as a transitional level whose deeper origin is associated with the (t_{-1}) pre-geometric process, while its observable gravitational manifestation belongs to the physical 4D universe.

Correspondence Among Dark Matter, the Light Cone, Baryonic Matter, and the Corrective Process

To conceptualize the relationship among **dark matter (DM), the light cone (LC), and baryonic matter (BM)** in the TLD model, a provisional philosophical–Buddhist correspondence may be introduced.

This correspondence does **not** imply that spiritual states are physically identical to cosmological structures. Rather, it is intended to illustrate different degrees of **manifestation, recognition, and correction**.

(i) Dark Matter — Minimal Manifestation and Minimal Correction

Within the TLD model, dark matter may be hypothesized as the physical manifestation associated with the minimal temporal level (t_1).

Symbolically, this state may be compared with structures or beings characterized by a relatively small degree of informational deviation or "karma," requiring only a minimal corrective process rather than complete manifestation within the photon–baryonic world.

At the opposite extreme, TLD may also hypothesize states in which informational deviation has become resistant to ordinary correction—symbolically corresponding to **Hell**. Such a correspondence is philosophical and does not imply that physical dark matter itself possesses karma or consciousness.

(ii) Light Cone and Baryonic Matter — Manifestation, Recognition, and Correction

When a structure is manifested at the (t_2) level, photons, baryonic matter, and the Lorentzian causal structure of 4D spacetime become fully relevant.

This is the domain in which informational deviation can become manifested as observable phenomena, interact with other structures, produce consequences, become recognized, and ultimately undergo correction.

In the symbolic language of TLD,

Biased information →t_2→photon+BM→ Corection

In a Buddhist philosophical correspondence, this may be regarded as the world in which **karma becomes manifested through actions and consequences**, thereby providing conditions for recognition, learning, transformation, and correction.

Nevertheless, **physical causality**, as represented by the light-cone structure, must be distinguished from **karmic causality** in Buddhist philosophy. The two may be placed in conceptual correspondence within TLD, but they should not be presented as the same physical mechanism.

(iii) The Liberated State and the Meaning of "Beyond the Light Cone"

For beings regarded as liberated—such as **Arhats or Bodhisattvas** in Buddhist traditions—a distinction should be made between the **physical body** and the **liberated or Buddha-nature dimension**.

While such a being remains alive in the phenomenal world, the physical body remains baryonic matter, interacts with photons, and remains fully subject to the causal structure of 4D spacetime:

$$\text{Physical Body} \longrightarrow \text{BM} \longrightarrow t_2 \longrightarrow \text{LC}_{4D} \longrightarrow \text{Physical Causality}$$

Liberation therefore does not mean that the physical body literally escapes the light cone. The body remains subject to physical laws and to the physical causal relations of the phenomenal world.

By contrast, within the philosophical interpretation of TLD, the **Buddha-nature or spiritual dimension of a liberated being** may be used as a correspondence for a state that is no longer bound by the ordinary process of karmic generation and karmic correction.In this sense, such a state may be described as

$$\text{Saints in 4D} \;\; \angle \begin{matrix} \nearrow \textit{body 4D} \\ \rightarrow \textit{beyond}\ \text{LC}_{4D\text{-}LTD} \end{matrix} \text{(outside LC [Buddhahood/Sainthood manifested] + compassionate activities)}$$

The spacelike region outside a Minkowski light cone remains part of ordinary relativistic 4D spacetime. It should therefore not be directly identified with a spiritual or timeless domain.

For this reason, TLD may introduce a distinct conceptual notation:

Here, **Beyond-LC$_{TLD}$** denotes a hypothetical broader domain comprising the ordinary 4D causal description together with the deeper TLD framework proposed to underlie it. In this construction, the spacelike region conventionally drawn outside the Minkowski light cone represents only one part of the 4D geometrical picture and should not itself be equated with

This correspondence allows TLD to relate physical manifestation, informational correction, and spiritual liberation while maintaining a necessary distinction between **physical causality, cosmological structure, and philosophical or karmic causality**.

CHAPTER V: FROM INFORMATION TO GRAVITY, AND ULTIMATELY TO MATTER

Summary

*According to the hypothesis proposed in this book, when **biased information** emerges within the **Timeless Domain (TLD)**, the original state of integrity and uniformity is no longer fully preserved. The difference between primordial information and biased information creates a need for stratification, ordering, and reorganization according to a before-and-after sequence. It is from this necessity that the concept of **time** begins to emerge.*

In this model, time is not a pre-existing entity. Rather, it represents a mechanism that enables the TLD to classify information, actualize it, and gradually transform it back toward the state of integrity. This process unfolds through three stages.

*At the first stage, when the degree of informational deviation remains very small and has not yet exceeded the threshold requiring correction, time exists only in the form of **proto-time**, denoted by* (t_0)*. At this stage, there is as yet no energy, field, or matter; time manifests only as the capacity to order and distinguish biased information from the pristine state of the TLD.*

When the degree of deviation exceeds a critical threshold, (\theta_{\min}), time is activated at a minimal level, denoted by (t_1) ***(minimal time)**. This is the stage at which information begins to be actualized within the **Interface Domain**, where pre-material structures emerge. According to the hypothesis of this model, **dark matter and the gravitational field belong to this stage**.*

As the degree of deviation continues to increase, the process of actualization advances to (t_2) ***(physical time)**, where quantum fields, elementary particles, and ordinary matter emerge within four-dimensional spacetime.*

The emergence of (t_1) *and* (t_2) *not only establishes different levels of stratification within reality but also causes deviations within informational structures to become increasingly manifest. From the perspective of this model, it is precisely this manifestation that creates the conditions through which information can undergo learning, correction, and gradual transformation back toward the state of integrity.*

Biased information is hypothesized to be describable by three fundamental geometric quantities:

- ***(\theta)**: the degree of informational deviation from the state of integrity;*
- ***(\kappa)**: informational curvature, hypothesized to be related to gravity and mass;*
- ***(\tau)**: informational torsion, hypothesized to be related to quantum spin and the degree of manifestation of material structures.*

*Within this model, **gravity is not a consequence of matter, but rather the first geometric manifestation of biased information during the process of temporalization**. Matter emerges only at a later stage, as a higher degree of actualization within the same underlying process.*

Matter, therefore, is not the ultimate purpose of existence, but rather a medium through which information manifests its own state. Material structures preserve the imprints of informational deviations, thereby providing the conditions for learning, self-correction, and transformation. In this sense, matter reflects the process of accumulation and transformation of information rather than constituting an ultimate, independently existing reality.

This perspective shares certain conceptual similarities with modern approaches proposing that gravity and the structure of spacetime may emerge from information or thermodynamic quantities. Examples include Ted Jacobson's approach, in which the Einstein field equations are derived from a thermodynamic relation, and the Unruh effect, in which temperature emerges in association with acceleration and causal horizons.

*In this conceptual sense, the model is compatible with approaches that investigate gravity and spacetime as potentially emerging from information or thermodynamics, particularly the work of Jacobson and the Unruh effect. **However, the TLD model is not***

identical to these theories. Instead, it proposes a different hypothesis: primordial structural information constitutes the foundation, while time, gravity, and matter represent successive stages in the actualization of a single underlying process.

*The TLD model therefore goes further by hypothesizing that **primordial structural information**, rather than matter, constitutes the primary foundation of the process of actualization.*

*At the philosophical level, this perspective also bears a conceptual correspondence to the Buddhist principle **"All phenomena are created by mind"** (Nhất thiết duy tâm tạo), according to which the phenomenal world is not ultimate reality but rather the manifestation of a deeper foundation. In the TLD model, this deeper foundation is hypothesized to be the domain of primordial structural information, from which **time, gravity, and matter emerge successively as consecutive stages of the same process of actualization**.*

If Chapter III presents time as an emergent mechanism through which information undergoes self-correction, we may now take one further step: once information has become biased, it not only requires time, but also gives rise to a tendency to return toward order. In the 4D world, this tendency is manifested as **gravity**.

The next question is: **How does time emerge, and why is time associated with dark matter different from time associated with ordinary matter?**

I. THE EMERGENCE OF TIME:

: From Proto-Time to the Temporalization of Information

As presented in Chapter III, in the CPT/TLD model, time does not emerge immediately in its fully developed form. Rather, it is actualized at different levels corresponding to the degree of informational deviation. This process may be represented schematically as:

Proto-time (t_0) → Minimal time (t_1) → Full time (t_2).

1. PROTO-TIME

In the primordial state of CPT, physical time does not yet exist. However, when CPT begins to differentiate into a domain of information in integrity and a domain of biased information, the need to organize informational structures also begins to emerge. This need does not yet generate a flow of time, but it creates the **possibility of ordering**. This state is called **proto-time ((t_0))**.

Proto-time is not yet time in the sense that it could be measured by a clock. It produces no motion, no energy, and no matter. It represents only the capacity to distinguish and order different informational states. Proto-time may therefore be understood as a latent principle of order: there is not yet a clearly defined "before" and "after," but there is already the possibility for one structure to be ordered differently from another.

Within this process of establishing order, the need to correct informational deviation becomes almost inevitable. Correction acts to reduce the degree of deviation, thereby allowing the ordering process within CPT/TLD to proceed more effectively.

However, before a deviation can be corrected, it must first be made manifest. This need for manifestation leads to the process of **actualization**, in which informational deviation becomes materialized. In other words, what is deviated within the invisible domain of structural information must become expressed as observable phenomena so that it can be recognized, measured, and transformed.

In this process, time functions as a principle for ordering and actualizing information. Through time, structural information from the invisible domain can be translated into the visible world through physical laws that science has discovered over the past several centuries, such as velocity,

$v = d/t$

acceleration, and the equivalence of mass and energy,

$E = mc^2$)

Therefore, as the degree of informational deviation increases, represented by the deviation angle (\theta), the difference between the state of integrity and the deviated state becomes increasingly explicit. At this point, the need for self-correction is no longer merely a latent possibility but becomes an actual process. It is here that time begins to be actualized in its first operative form.

2. TIME (t_1)

This level of time is called **minimal time ((t_1))**.

It is an extremely weak form of time. It is not sufficient to generate the familiar material world, but it is sufficient to allow **informational curvature** to manifest as **gravity**. At the same time, it permits a primitive form of matter to emerge.

According to the TLD hypothesis, this first form of matter is **dark matter**.

Dark matter does not emit light, appears to interact little or not at all through ordinary electromagnetic processes, and remains extremely difficult to detect directly with current physical methods. Nevertheless, its gravitational effects are evident through phenomena such as the motions of galaxies and gravitational lensing.

This suggests a noteworthy possibility. According to the HTTV model, if gravity emerges before ordinary matter, then dark matter may represent the first stage in the actualization of information.

At this stage, information has deviated sufficiently to manifest gravitational effects, but not sufficiently to generate the complete set of electrically charged particles, atoms, and molecules characteristic of the baryonic world.

From this perspective, dark matter is not an entirely independent form of matter unrelated to ordinary matter. Rather, it represents **a lower level of actualization of the same underlying informational structure**, corresponding to the (t_1) temporal regime.

Dark matter, in this model, is not directly related to dark energy.

3. TIME (t_2)

As the actualization of time proceeds further, **minimal time (t_1) does not gradually transform into full time (t_2)**. Instead, according to the CPT/TLD hypothesis, **(t_2) is activated directly from (t_0) within CPT**, just as (t_1) represents another level of temporal activation.

Thus,

.The (t_2) regime corresponds to the physical time we experience in everyday life and to the time appearing in classical mechanics, quantum mechanics, and relativity. With (t_2), information manifests not only through gravity but is further actualized as energy, quantum fields, elementary particles, atoms, molecules, astronomical bodies, and ultimately life.

Accordingly, ordinary matter is not the starting point of the process of creation, but rather a more developed stage in the actualization of information. Time does not emerge once and then remain unchanged; instead, different temporal regimes accompany different levels of informational actualization.

According to this hypothesis, dark matter and ordinary matter are therefore not two completely unrelated realities. They reflect two different levels of the same underlying process, distinguished primarily by the degree and mode through which time is actualized.

Within the TLD model, the progression from information to the phenomenal world may therefore be summarized as:

Canonical information → biased info → Curvature → Gravity → Proto-time (t_0) → Minimal time (t_1) → Dark Mater .
→ Full time (t_2) → Baryonic Matter → Life

From this perspective, time is not merely a dimension of the universe. **Time is the bridge between a timeless informational foundation and the visible material world.** Through different levels of temporal actualization, the invisible gradually becomes visible, allowing the process of informational self-correction to unfold within the four-dimensional universe.

In summary, according to the CPT/TLD model, throughout the **temporalization** of structural information that gives rise to a living individual in the material world, the **proto-individual is not transformed into another entity**. It continues to exist as the underlying informational structure within CPT. What becomes actualized through the Interface Domain and the 4D world is only the **temporalized version** of the proto-individual.

In other words, the proto-individual is the **prototype**, whereas the individual in the 4D world is the **phenotypic manifestation** of that prototype under the influence of time, matter, and biological conditions.

The proto-individual continuously maintains a connection with the individual in the 4D world through the Interface Domain. This relationship, however, is not completely symmetrical. From the CPT side, the proto-individual may continuously influence the individual through intuition, innate tendencies, inspiration, inner awareness, and processes involved in the formation of knowledge. Conversely, from the perspective of the 4D individual, access back to the proto-individual is restricted by the orthogonality between CPT and spacetime, as well as by the limitations imposed by the senses, the brain, and linear time.

This relationship may be illustrated through the analogy of a **playwright, a script, a director, and a stage performance**.

The playwright creates the script, corresponding to **structural information within CPT**. The director translates the script into an unfolding dramatic sequence, corresponding to **time and temporalization**. The play is then performed on the stage, corresponding to **phenomena manifested in the 4D world**.

The actor remains dependent upon the pre-existing script. The actor's actions unfold according to the structure established by the playwright. Yet the actor, as a character within the play, does not directly act upon the playwright who created the script.
Only when the perspective is shifted beyond the internal framework of the play can the observer recognize that the deeper source underlying the actor and the performance is the playwright and the script.

Another Analogy

Suppose a composer conceives a symphony. In the composer's mind, the entire symphony already exists as an integrated whole. When the composer writes it on paper, however, the music must be expressed through staffs, notes, durations, rhythm, and other musical symbols.
Those symbols do not create the symphony. They merely represent something that already exists at a deeper conceptual level.
In this analogy, the audible music does not contain the totality of the composer's original conception; rather, it is the **manifested expression** of that conception under the constraints of notation, instruments, sequence, and time. Likewise, in the CPT/TLD model, the physical universe may be understood as the temporally ordered and materially actualized manifestation of a deeper domain of structural information.

DARK MATTER AND ORDINARY MATTER

In the TLD model, **dark matter is not regarded as matter that has failed to become baryonic matter**, nor is it necessarily a state that will eventually transform into ordinary matter. Instead, dark matter is hypothesized to consist of informational structures that have reached the state of **minimal temporalization t1** but have not, or do not, proceed to the state of **full temporalization t_2**.
The boundary between dark matter and baryonic matter is therefore determined not only by the degree of informational deviation, but primarily by **the capacity of the informational structure itself to transform and reorganize**. Only those informational structures capable of further evolution, while simultaneously satisfying the conditions of informational geometry—the anglem θ curvature κ and torsion —τ as well as the

symmetry requirements associated with the Standard Model, become manifested as baryonic matter. The remaining structures continue to exist in the dark-matter state.

In other words, **the relationship between (t1) and (t_2) is not a continuous transition but proceeds through distinct levels or regimes**. This interpretation is based on the observation that dark matter, hypothesized in this model to belong to the (t_1) domain, and ordinary matter, associated with the (t_2) domain, do not ordinarily exhibit an observable continuous transitional state between the two forms.

Although there are hypotheses suggesting that **dark matter may contain, or be associated with, hypothetical particles capable of interacting or converting into ordinary particles**, such possibilities remain speculative and have not been experimentally established.

From this perspective, dark matter and baryonic matter are not necessarily two completely separate categories of matter. Rather, they may represent two different states of manifestation arising from the same underlying informational and temporal foundation. If transitions between these states occur in particular circumstances, they would not represent a simple direct conversion of one type of matter into another. Instead, they would result from a **transformation of the underlying informational structure**, altering the degree of temporalization and the manner in which information manifests within the physical world.

Within the ontological interpretation of the TLD model, informational domains of very high density but extremely low transformational capacity may be symbolically associated with the image of **"hell"** in Buddhist thought. This correspondence is philosophical and is intended only to represent different degrees of informational transformability. It does **not** imply that dark matter in general corresponds to a religious concept of hell.

II. GRAVITY AND MATTER: REVERSING THE CONVENTIONAL DIRECTION

In modern physics, gravity is commonly understood through two major frameworks. In Newtonian mechanics, gravity is an attractive interaction between bodies possessing mass. In Einstein's general relativity, gravity is no longer treated as a force in the ordinary Newtonian sense; instead, freely falling bodies follow the geometry of spacetime, whose curvature is related to the distribution of matter and energy.

Einstein's field equation may be written schematically as

$$R_{\mu\nu} - \frac{1}{2} R g_{\mu\nu} + \Lambda g_{\mu\nu} = \left(\frac{8\pi G}{c^4}\right) T_{\mu\nu}$$

where ($G\mu\nu$ represents the Einstein tensor describing spacetime curvature, ($g\mu\nu$ is the metric tensor, λ is the cosmological constant, (G) is Newton's gravitational constant, (c) is the speed of light, and ($T\mu\nu$) is the stress–energy tensor.

A. MATTER IS NOT THE ULTIMATE REALITY

The CPT/TLD model, however, proposes a deeper reversal of this conventional direction. Matter is not regarded as the ultimate cause from which gravity originates. Instead, the deeper source is hypothesized to be **informational deviation within CPT**. Matter appears only as a later manifestation of this process.

In other words, matter does not first curve the underlying informational structure. Rather, according to the TLD hypothesis,

Biased Information ⟶ Informational Curvature ⟶ Gravity ⟶ Weight/Mass Manifestation ⟶ Material Phenomena}

In the state of integrity, information within CPT is balanced, timeless, and requires no corrective process. When biased information appears, however, this equilibrium deviates from its primordial axis of integrity. Such deviation may be represented geometrically as a curvature of informational structure.

This curvature is **not yet Einsteinian spacetime curvature**, because physical spacetime has not yet fully emerged at this stage. It is instead hypothesized to be a form of **pre-physical curvature**, or curvature of the underlying informational structure.

When informational curvature becomes actualized within the Interface Domain between CPT and the 4D world, it manifests as gravity. In this sense, gravity is not merely a primitive force pulling one object toward another. Within the TLD interpretation, **gravity is the geometric sign that information has departed from its equilibrium state and carries a tendency toward the restoration of order**.

This expresses the principle that "the invisible gives rise to the visible." Biased information itself is invisible. Gravity is likewise not directly visible; its effects are observed. Through gravity, material structures acquire measurable gravitational properties; through measurement, the 4D world becomes characterized by form, position, interaction, and history.

The proposed causal direction therefore differs from the conventional interpretation in which matter–energy sources spacetime curvature.

Within TLD, gravitational properties are not merely attributes possessed by already-existing matter. They are visible manifestations of a deeper condition: **deviation and curvature within the timeless informational foundation**. Matter exhibits gravitational properties because it carries the manifestation of this deeper informational curvature.

Matter, therefore, is not ultimate reality. It is a manifestation through which biased information can become explicit enough to be recognized, interacted with, and corrected.

If information remained completely in integrity, there would be no ne ed for materialization. There would be no need for space, physical time, motion, or weight. But once deviation appears, according to this model, a domain is required in which that deviation can become manifest. **That domain is the 4D world.**

From this perspective, the material universe does not exist primarily for the comfort, pleasure, or entertainment of sentient beings. Such experiences are secondary phenomena within the 4D realm. The deeper function of material manifestation is to make informational deviations recognizable, thereby creating the possibility of self-correction and restoration toward the original state of integrity.

Gravity therefore acquires a special significance in the TLD framework. It is not merely a physical interaction but the first geometric trace of deviation from equilibrium within the informational foundation.

Thus:

Informational Deviation→Informational Curvature→ Gravity→Material Manifestation→Recognition and Correction

Where gravity appears, there is evidence of curvature at the physical level. Within the additional hypothesis of TLD, such physical curvature reflects a deeper informational curvature. Where informational curvature exists, there is an informational structure that has departed from complete integrity; and where such deviation exists, there arises the need for time, manifestation, and a corrective process.

In summary, within the CPT/TLD model, gravity acts as a bridge between the invisible and the visible. It connects biased information in CPT with gravitational and material phenomena in the 4D world.

This offers a different interpretation of gravity: it is not regarded merely as attraction, nor solely as spacetime curvature, but as the **physical geometric manifestation of a deeper informational curvature along the pathway from CPT into the phenomenal world.**

Philosophical and Religious Reflection: Connections with Religious Traditions

If the TLD model is extended into the philosophical and religious domain, **the curvature of 4D spacetime may be interpreted, at the symbolic and ontological level of the model, as the physical manifestation of a deeper curvature within informational structure**.

According to this hypothesis, the further information deviates from its primordial state of integrity, the greater the corresponding deformation of its informational geometry. When temporalized and actualized, this deformation may become manifested as gravitational curvature, mass, and material structure.

At the ethical and spiritual level, **informational deviation** may be compared—without identifying these concepts as equivalent—with different ideas found in religious traditions: **unwholesome karma** in Buddhism, **sin** in Christianity, or departure from the guidance and will of Allah in Islam.

These concepts are not theologically identical. Nevertheless, they share a broad theme: human beings may depart from an order or standard that must subsequently be recognized and addressed.

Under the TLD interpretation, the material world may therefore be conceived as **a domain in which deviations that remain abstract at the informational level become manifested in more concrete forms**. In this symbolic sense, materialization is not the "punishment" of deviation. Rather, it is a process through which deviation becomes recognizable, experienceable, and transformable.

This perspective has a notable point of correspondence with Buddhism, Christianity, and Islam: none of these traditions regards present worldly existence as the ultimate or eternal state of human existence. Although their teachings concerning the purpose of life differ substantially, all place importance on forms of **trial, learning, moral cultivation, correction, or transformation**, while orienting human existence toward a condition beyond ordinary worldly life.

In the Islamic tradition, a well-known hadith attributed to the Prophet Muhammad states:

"The world is a prison for the believer and a paradise for the unbeliever."

The meaning is not that the physical universe is literally a prison created by the geometry of spacetime. Rather, the believer lives within moral limitations, responsibilities, and disciplines, while worldly existence itself is not regarded as the ultimate end.

FROM TLD TO YANG–MILLS THEORY: A HYPOTHESIS OF INFORMATION, TIME, AND THE MANIFESTATION OF REALITY123

Yang–Mills theory provides the mathematical foundation for describing fundamental interactions through the concepts of **gauge symmetry** and **gauge fields**. Like general relativity and quantum mechanics, however, it primarily describes structures operating within physical spacetime rather than providing an established account of why spacetime, gauge symmetries, or the particular physical fields exist in the first place.

Yang–Mills and the Limits of Physical Description

Yang–Mills theory demonstrates how fundamental interactions can be formulated through local gauge symmetry. Gauge fields arise mathematically when local gauge invariance is imposed, and the success of gauge theories within the Standard Model represents one of the major achievements of twentieth-century physics.

Yet a deeper question may still be asked:

Why does nature possess these particular gauge structures?

Yang–Mills theory describes fields and particles within its domain of applicability. By itself, it does not provide an ultimate ontological explanation for why the physical gauge structure takes the form

$SU(3)_C \times SU(2)_L \times U(1)_Y$

or why spacetime exists as the arena in which these fields operate.

This leaves open the possibility—at the speculative level—that a deeper layer of reality could underlie both gauge fields and spacetime itself.

The Significance of Yang–Mills Theory in the TLD Model

One of the most important achievements of modern physics is the recognition that fundamental interactions can be formulated through the principle of **local gauge symmetry**.

Within the Standard Model, the gauge structure

$SU(3)_C \times SU(2)_L \times U(1)_Y$

underlies the strong and electroweak interactions.

Historically and mathematically, the essential point is not that "entities can exist only when symmetry exists." Rather, gauge symmetry provides the structural framework from which the corresponding gauge fields and interactions are formulated.

In the TLD model proposed in this book, however, gauge symmetry is not regarded as the ultimate starting point. Instead, it is hypothesized to be **the physical manifestation of a deeper level of informational order**.

The model proposes a philosophical extension in which physical structures arise from a deeper pre-temporal informational domain called the **Timeless Domain (TLD)**.

Within this framework, primordial mathematical relationships constitute a form of pregeometry, while information is divided into two broad classes: **non-string information** and **string-like information**. String-like information is not itself matter; it represents informational structure requiring temporalization before it can become physically manifested.

M-theory may then be used as a conceptual analogy for an interface between the TLD and the physical world, with brane-like structures serving as possible mathematical representations of informational organization, while time provides the mechanism through which information becomes physically manifested. This interpretation is an extension proposed by the TLD model and should not be identified with M-theory itself.

The TLD is hypothesized to be a primordial informational domain in which information initially exists in a state of integrity, prior to physical stratification and prior to the emergence of spacetime. In that state, information possesses its highest degree of consistency and symmetry.

As actualization begins, these primordial symmetries are hypothesized to become progressively expressed through the Interface Domain and eventually appear within spacetime as the gauge symmetries described by Yang–Mills theory.

Accordingly, the groups (U(1)), (SU(2)), and (SU(3)) would not constitute the ultimate ontology of reality. Within the TLD hypothesis, they would instead represent **mathematical manifestations, within the physical world, of deeper informational regularities.**

Thus, in the IPTM framework, gauge symmetry may be interpreted as a secondary, localized stage in the broader sequence:

Information→Informational Geometry →Temporalization →Gauge Structur→Quantum Fields →Particles and Matter

INFORMATION, THERMODYNAMICS, AND GRAVITY

Research on black holes revealed a remarkable conjunction of **gravity, entropy, and temperature**, encouraging physicists to investigate whether gravity might be connected to deeper thermodynamic and informational principles.

Ted Jacobson showed that, under appropriate local-horizon assumptions, the Einstein field equation can be obtained from a thermodynamic relation of the form $\boxed{\delta Q = T\,dS}$

where:

δQ=heat flow,T=temperature,dS=change in entropy.

This suggests that Einstein's equation may admit a thermodynamic interpretation rather than necessarily being fundamental at every microscopic level—somewhat as macroscopic equations of state arise from deeper microscopic degrees of freedom.

Within the language of the TLD model, this result is particularly suggestive. If information in integrity is regarded as a state of maximal informational clarity or "luminosity," entropy may be interpreted within the model as a measure of information becoming obscured or deviating from its reference state.

A black hole represents an extreme physical regime in which spacetime curvature and gravitational effects become enormous, horizon entropy becomes extremely large, and light originating inside the event horizon cannot escape to infinity.

From the perspective proposed here, a black hole may therefore be interpreted not merely as an astronomical object, but as a physical setting in which **geometry, thermodynamics, and information become exceptionally closely interconnected**.

This motivates a deeper question: perhaps gravity, time, and even spacetime itself are manifestations of structures whose underlying description is informational.

Informational Curvature and the Graviton

Within the TLD model, gravity is not assumed to originate fundamentally from the exchange of gravitons. Instead, gravity is hypothesized to arise macroscopically from curvature within an underlying informational structure.

Curved informational fibers represent deviations from the primordial equilibrium state. If such structures admit quantum excitations, those excitations could, within the model, appear as **graviton-like modes**.

Thus, if gravitons exist, the TLD interpretation reverses

TLD→Structural Inform→Curved Fiber→Gravity→Graviton

The graviton would therefore not create the underlying gravitational structure. Instead, it would be the quantum manifestation of an already existing deeper geometry.

An analogy is provided by waves on the ocean: **the waves do not create the ocean; they reveal the dynamics of the ocean**. Similarly, within the TLD hypothesis, the graviton would reveal the quantum dynamics of gravitational geometry, while gravitational geometry itself would reflect deeper informational curvature.

In the conventional description of general relativity, stress–energy and spacetime geometry are related through Einstein's field equation. The TLD model proposes an additional, deeper layer beneath this established physical relationship:

κinformation→Gravitational Geometry↔Matter–Energy

The proposal has conceptual affinities with research programs associated with Jacobson and with broader attempts to connect gravity, thermodynamics, and information. It should nevertheless be distinguished from those established results and proposals: **TLD specifically hypothesizes that structural informational curvature precedes gravitational geometry**.

If TLD constitutes a pre-physical and pre-spatiotemporal foundation, gravity would no longer represent the first ontological cause. It would instead appear when informational structures depart from primordial equilibrium and their curvature becomes physically manifested.

Accordingly, the graviton—if it exists—would not be the entity that "creates" gravity, but a quantum excitation associated with an already curved gravitational structure.

Gravity, in this view, becomes **the geometric shadow of a deeper informational layer**.

Just as temperature is a macroscopic variable related to microscopic degrees of freedom, gravity might be interpreted as the macroscopic manifestation of organization and curvature within a deeper informational structure.

From the CPT/TLD perspective, when information departs from primordial integrity, it becomes biased, giving rise to curvature and torsion. These deformations create a need for self-correction aimed at restoring informational order. The mechanism enabling this corrective process is manifested through the emergence of time.

Depending on the degree and character of the deviation, temporalization may occur at different levels. For deviations that require only limited physical manifestation—or for structures whose capacity for transformation is restricted—**minimal time (t_1)** appears only weakly within the Interface Domain, allowing dark matter to exist as an intermediate regime of material manifestation.

Conversely, when informational structures contain sufficient curvature, torsion, and complexity to require extensive manifestation, interaction, and transformation, **full physical time (t2)** is activated. Together with physical spacetime, biased information becomes geometrically expressed through fields and material structures.

Within this interpretation, **Wick rotation** may be used as a mathematical analogy for the transition between a regime without ordinary physical time and the Lorentzian spacetime of the phenomenal world. It should not necessarily be interpreted as a literal physical process occurring in nature; rather, it is a mathematical transformation that helps illustrate a change between different geometric regimes.

The "knots" within informational structure become manifested in the material world as limitations, conflicts, suffering, and consequences of actions. Once exposed within time, sentient beings can recognize these deviations, learn from their consequences, and gradually modify themselves.

In this philosophical sense, time is not merely a measure of change. It is also the domain within which karma can become manifested, recognized, and transformed.

III. TEMPERATURE AND HORIZONS

The discoveries associated with black-hole thermodynamics and quantum field theory in curved spacetime revealed unexpected relationships among **horizons, temperature, entropy, and acceleration**.

The Unruh effect, in particular, predicts that a uniformly accelerated observer in the Minkowski vacuum perceives a thermal spectrum, with an associated temperature

$T_U = \hbar a / 2\pi c k_B$

This relationship is remarkable because it connects acceleration and quantum fields with a temperature perceived by the accelerated observer.

From the perspective of the TLD hypothesis, such relationships suggest that thermodynamic quantities need not be viewed simply as properties of conventional bulk matter. They may instead provide clues to deeper connections among geometry, horizons, quantum information, and physical observation.

The TLD model extends this idea speculatively by proposing that **entropic effects within the Interface Domain could arise from deformation of informational structures as they begin to couple to time**, potentially before fully developed baryonic matter emerges.

JACOBSON'S ARGUMENT (1995)

Jacobson considered the thermodynamic relation $\boxed{\delta Q = T dS}$

By applying this relation locally to causal horizons, together with the proportionality between entropy and horizon area and the Unruh temperature, he showed that the Einstein field equation can be recovered in the appropriate form: $\boxed{8\pi G T_{\mu\nu} / c^4}$

The significance of this result is profound. It establishes a deep relationship between **spacetime geometry and thermodynamics** and raises the possibility that Einstein's gravitational field equation may behave like an equation of state arising from deeper microscopic degrees of freedom.

For the TLD framework, this provides an important conceptual analogy—but not experimental proof of TLD. The proposed distinction can be summarized as:

- Jacobson:Horizon Thermodynamics→Einstein Gravitational Dynamic

Similarly:

- TLD Hypothesis:Structural Information → Informational Curvature → Temporalization → Gravitational Geometry → Matter .

Thus, Jacobson's result does not demonstrate that structural information in the TLD sense generates gravity. Rather, it provides a powerful example showing that **gravity, geometry, entropy, and thermodynamics may be more deeply interconnected than a purely matter-first description would suggest**.

FROM THE PRINCIPLE "ALL PHENOMENA ARE CREATED BY MIND" TO THE TLD MODEL

A Hypothesis on the Process of Material Actualization

In Buddhist thought, the principle **"All phenomena are created by mind"** (*Nhất thiết duy tâm tạo*) carries not only ethical or psychological meaning, but also suggests a deeper cosmological principle: **the visible emerges from the invisible**.

This idea is consistent with the spirit of Prajñāpāramitā expressed in the statement:

"Form is Emptiness; Emptiness is Form."

Matter is therefore not regarded as ultimate reality, but rather as the manifestation of a deeper foundation.

The TLD model (**Timeless Domain / Transcendent Logic Domain**) hypothesizes that this deeper foundation is a domain of **Intrinsic Structural Information**, denoted by (I_s). Within this domain, matter, energy, and physical time do not yet exist. What exists instead are structural relations, organizing principles, and potentials for manifestation.

According to the TLD hypothesis, the process of cosmic manifestation does not begin with matter. It begins with structural information operating through the **Interface Domain**, where the transition toward the 4D world begins and where time first becomes activated, allowing forces and matter to emerge.

The process may be represented schematically as

Mn∝IsGnTnHδn2,n=0,1,2

This may be read conceptually as:
Mass is a function of structural information, gravitational manifestation, the temporal regime, and the Higgs field, with full Higgs participation occurring in the (n=2) phase.
Here, $H_0=0, H_1=0, H_{\delta n2}=H$for n=2
Thus, the model may be written schematically as
M= |0,t=0,TLD,
|DM,t=1,(thin time),
|BMn=Is G H,t=2,(full time). \
with threshold conditions represented schematically by the gravitational or informational activation levels associated with each temporal regime.
The three regimes are interpreted as follows:

- **(t_0)**: pure TLD; no physical time and no matter;
- **($_{t1}$)**: gravity has appeared together with a thin or minimal form of time, corresponding to the dark-matter domain or an intermediate regime of manifestation;
- **(t_2)**: full physical time is present, the Higgs field participates fully, and baryonic matter becomes manifested.

Within this framework, gravity (G) is not assumed to be merely the final consequence of matter. Instead, it is proposed as an earlier physical condition associated with the emergence of time. When the relevant gravitational or informational threshold is crossed, time becomes activated. Only when time reaches the full (t_2) regime does the Higgs field become fully involved in the manifestation of ordinary visible matter.
From this perspective, matter is no longer the first cause of gravity. Rather, matter becomes the final stage of a longer sequence of actualization beginning from structural information.
This interpretation does not deny the experimental success of general relativity, which accurately describes the relationship between matter–energy and spacetime curvature. Instead, TLD asks whether that relationship represents the ultimate ontological cause, or whether it may itself be the physical manifestation of a deeper informational process.
If gravity appears prior to ordinary matter, and if the Higgs sector becomes fully expressed only when time reaches the (t_2) regime,

then matter may be regarded as a late stage in a sequence beginning with structural information.

At the philosophical level, this interpretation corresponds to the principle **"All phenomena are created by mind"**, while suggesting that the conventional question—"Does matter generate gravity?"—may need to be reconsidered at a deeper ontological level.

From this perspective, the relationship between matter and gravity may not be a simple one-directional causal relationship. Instead, both may be different manifestations of the same deeper process of actualization originating from primordial structural information.

Even if gravity is hypothesized to arise before matter, however, the established physical effect of gravity in curving spacetime remains unchanged. The equations and experimentally verified laws of general relativity remain applicable at the physical level. The TLD proposal concerns a deeper ontological ordering rather than a replacement of existing gravitational calculations.

Thus, gravity may be interpreted as **the first physical manifestation of structural information that has deviated from its state of primordial integrity**.

Within this symbolic framework, dark matter (DM) is not merely an unresolved physical phenomenon. In the language of TLD, it may also be interpreted as a form of **latent karmic structure that has not yet become fully manifested as visible matter**.

It exhibits gravity, but does not display the full range of manifestation through light, electromagnetic interaction, and ordinary Higgs-mediated mass generation characteristic of baryonic matter.

If temporalization and Higgs-mediated manifestation proceed further, some forms of biased structural information may, in principle within the model, become crystallized into ordinary matter.

Matter, therefore, is not merely "matter" in a neutral physical sense. At the symbolic and philosophical level of TLD, it may also be understood as **the visible trace of karmic structure manifested within spacetime**.

From this perspective, the visible world becomes a **school of karma**.
Matter, the body, circumstances, and gravity are not interpreted as realities entirely separate from consciousness, but as different levels of manifestation of structural information that has entered time.
The principle **"All phenomena are created by mind"** therefore does not merely mean that mind constructs subjective experience. Within the broader metaphysical interpretation proposed by TLD, it also suggests that the entire phenomenal order may represent the manifestation of informational structures that have either been conditioned by ignorance or progressively purified through wisdom.
The conceptual sequence may therefore be summarized as

Primordial Structural Information
→Informational Deviation→Informational Curvature→Gravity→Temporalization→|t_1→Dark Matter
|t_2→ Higgs Participation→Baryonic Matter.

At the physical level, general relativity continues to describe the relationship between matter–energy and spacetime curvature. At the ontological level proposed by TLD, however, both matter and gravitational geometry are interpreted as downstream manifestations of a deeper informational structure.

CHAPTER VI: THE INTERFACE DOMAIN (I): KARMIC CONSCIOUSNESS OR THE SOUL

Summary

This chapter presents the initial stage of the stratification process from the Timeless Domain (TLD) into the Interface Domain, where time begins to emerge in the form of ***minimal time*** t_1*). According to the proposed model, the Interface Domain serves as a bridge between TLD and the 4D material world, while also being the domain in which* ***Karmic Consciousness (Soul)*** *exists. The Soul is not regarded as an entirely immaterial entity, but as a subtle information–matter structure possessing characteristics of both domains. Because of its subtle nature, the Soul cannot be directly observed, but it may have a measurable weight. The portion oriented toward TLD retains the nature of timeless information, while the portion oriented toward 4D begins to acquire characteristics of* ***Dark Matter*** *and may be associated with hypothetical particles such as sterile neutrinos, axions, or dark photons. These components currently remain only theoretical hypotheses and have not been experimentally confirmed.*

In the TLD model, the Soul serves as an intermediary medium for transmitting, storing, and transforming information between TLD and the brain. The brain encodes information from the 4D world, while the Soul stores the informational structure of karmic consciousness and maintains the continuity of memory, personal identity, and the learning process. The interaction between the Soul and the nervous system is hypothesized to occur through synapses and very weakly interacting particles belonging to the Interface Domain.

The model also proposes a distinction among three concepts: ***information***, ***knowledge***, *and* ***awareness***. *Information is merely structure or code; knowledge emerges when information is interpreted by the proto-individual (the Pre-Individual in TLD/TLD); whereas Primordial Awareness corresponds to the state of integral information in TLD and may be compared with* ***Prajñā Wisdom*** *in Buddhist terminology. From this perspective, enlightenment is not the accumulation of additional knowledge, but the process of gradually removing temporalized and deviated* *layers of information so that Primordial Awareness may reveal itself.*

The proto-individual is regarded as the primordial informational structure of each individual, existing before the biological body. When the process of temporalization occurs, the proto-individual enters the Interface Domain, forms the Soul, and then combines with the brain and body to manifest as a living individual in the 4D world. According to this hypothesis, parents do not create the proto-individual but only provide the biological environment through which that informational structure becomes actualized. According to the TLDmodel, the exchange of information between the timeless domain and the biological brain is ***asymmetric****.*

According to this hypothesis, the relationship between TLDand the brain is not a completely symmetrical two-way exchange of information. Instead, it is a mechanism in which ***the receiving direction is nearly complete, whereas the direction of manifestation is selective****, in order to prevent information overload and ensure that only structures meaningful to the processes of cognition and development are actualized in the 4D world.*

From the theoretical framework above, the TLD model also proposes several hypotheses for interpreting phenomena such as the formation and retrieval of memory, the continuity of karmic consciousness, Near-Death Experiences (NDE), Out-of-Body Experiences (OBE), and certain special states of consciousness.

According to this hypothesis, the "Soul," possessing the nonlocal characteristics of the Interface Domain, is hypothesized to interact with the nervous system through changes in electrical potential at synapses and neuronal terminals. When the electrical-potential state of the neural network changes, the degree of connection between the Soul and the brain may change correspondingly. Under special conditions, such as deep meditation, near-death experiences, or certain other states of consciousness, the model hypothesizes that this connection may temporarily weaken, producing phenomena commonly described as out-of-body experiences.

These hypotheses have not currently been confirmed by modern physics or neuroscience. They are proposed as a direction for interdisciplinary research aimed at developing models that may be testable in the future, connecting theoretical physics, neuroscience, and Buddhist philosophy on the foundation of a unified informational model.

According to the **TLD (Timeless Domain)** hypothesis, TLD is a foundational domain of reality that human beings are almost unable to perceive directly through the senses or through current methods of physical measurement. In contrast, the 4D world is the phenomenal domain that science has studied for centuries, discovering postulates, laws, and mathematical models describing the operation of matter, energy, space, and time.

Although very different in nature, TLD and the 4D world are not two separate realities, but two continuous domains of the same process of actualization. It is precisely this continuity that gives rise to an **Interface Domain**—the transitional region between the timeless reality of TLD and the spacetime reality of the physical world. This domain arises from the principle of uninterrupted continuity, necessarily connecting the two domains into a single reality.

The Interface Domain does not belong entirely to TLD, but neither is it yet entirely the 4D world. Here, characteristics of both domains coexist. On the one hand, structural information still retains its foundational nature and has not yet been fully materialized, as in TLD; on the other hand, the first manifestations of time, informational curvature, gravity, and matter have begun to appear. Therefore, the Interface Domain may be regarded as the place where the transformation from the invisible to the visible, from potentiality to actuality, and from pure information to physical phenomena takes place.

According to the TLD model proposed in this book, many difficult-to-explain phenomena of modern physics may originate precisely within this Interface Domain. This may be where **"thin time"** t_1, **proto-gravity**, **proto-dark matter (DM_0)**, and **pre-geometric structures** appear before being fully actualized in 4D spacetime. Therefore, the Interface Domain serves not only as a bridge between the two domains of reality, but also as a key to understanding the process of material actualization and the transition from structural information to the phenomenal world.

In addition, dark matter is known to produce gravitational effects, although its nature has not yet been identified. In many theoretical physics models, dark matter has been hypothesized to be related to supersymmetric particles or other new particles beyond the

Standard Model. However, the possibility cannot be excluded that dark matter is not composed of material particles in the ordinary sense, but instead reflects geometric or informational properties of a deeper intermediate domain.

From the perspective of the TLD model, this suggests that the **Interface Domain** may be an appropriate place for interpreting the origin of gravity and dark matter. At the same time, from a philosophical perspective, the **Five Aggregates**—which are important components constituting the mental life and karmic consciousness of each individual—appear to have several structural correspondences with the additional dimensions from D6 to D11 in M-theory, although these two concepts are not completely identical.

According to the hypothesis proposed in this book, this correspondence is not intended to identify Buddhism with M-theory, but merely to suggest that the additional dimensions D6D6–D11D11 may be regarded as a possible mathematical representation of the portion of the Interface Domain **approaching the four-dimensional world**. According to this hypothesis, many phenomena commonly attributed to additional dimensions—such as gravity, dark matter, and pre-material structures—may be understood as manifestations of the Interface Domain during the process by which primordial information is transformed into the material world.

A. SOUL OR KARMIC CONSCIOUSNESS (KARMIC CONSCIOUSNESS / SOUL)

If TLD and the 4D material world (or 4D as described within multidimensional theories such as 11D/M-theory) are indeed two domains of different natures that nevertheless interact with one another, then, according to the model proposed in this book, the existence of an **Interface Domain** is a logical consequence. This domain serves as a bridge, allowing information to be transformed and transmitted between the timeless informational domain and the material world of spacetime.

According to the hypothesis of the TLD model, the foundation of the Interface Domain is manifested through **Karmic Consciousness**, also called the **Soul**. Karmic Consciousness does not belong entirely to TLD, but neither does it yet belong entirely to the material world. It is an intermediate informational structure possessing characteristics of both domains, thereby enabling it to maintain the continuity of information between TLD and the brain.

In this model, Karmic Consciousness not only reflects the mental state of a present life but is also hypothesized to preserve the entire informational structure of each individual. This structure includes imprints formed during the present life as well as information accumulated from previous lives, if their existence is real. Therefore, Karmic Consciousness is regarded as a **continuous informational structure**, ensuring continuity in the process of learning, transformation, and development of the **proto-individual** throughout multiple stages of actualization.

From this perspective, the Soul is not a supernatural entity separated from the laws of nature, but an intermediate component of the process of informational transformation. It is through Karmic Consciousness that the informational structures of TLD may gradually be manifested in the material world, while the experiences and learning acquired during physical life are simultaneously fed back to the primordial informational foundation.

As presented in previous chapters, the TLD model proposes a new perspective on the foundation of spacetime. According to this model, the four-dimensional world (4D) is not primordial reality, but an **"educational environment,"** in which deviated informational structures are stratified, revealed, and transformed. The emergence of spacetime therefore carries not only physical significance but also a functional meaning for the entire process of informational evolution.

From this perspective, time is not merely a coordinate axis in physics. Time is the mechanism that allows three fundamental processes to occur:

1. **Classification and stratification of information**, so that different structures can be identified and separated.
2. **Revelation of deviated information**, allowing deviations that remain latent to become apparent through processes of motion and interaction.

3. **Correction and transformation of information**, including:
   - (a) correction directly within the TLD domain;
   - (b) actualization into the 4D world as an environment for comprehensive learning and transformation.

According to this hypothesis, the emergence of time may proceed through several levels.

At the initial stage (**proto-time,** t_{01}, time exists only as a latent possibility, sufficient to establish the ordering and stratification of information but not yet forming a physical flow of time.

At the next stage (**minimal time,** t_1), time becomes more clearly defined, allowing deviated information to become manifest and to persist in a semi-material state. At this stage, the amount of time is still very small, insufficient to form ordinary matter or the rich energetic processes found in the 4D world.

If this hypothesis is correct, one noteworthy possibility is that the combination of deviated information and minimal time may manifest as **Dark Matter**. From this perspective, Dark Matter is not merely an unknown physical component, but also an intermediate state between the timeless informational domain and the material world. It still possesses gravitational properties but interacts very weakly electromagnetically, consistent with many observed characteristics of dark matter.

This hypothesis also suggests another interpretation of why Dark Matter constitutes approximately 27% of the total mass–energy content of the universe. Rather than viewing it merely as a "missing" material component, the TLD model asks whether this large proportion might reflect the very process of informational transformation before ordinary matter is formed.

If so, the Interface Domain between TLD and spacetime may be the place where many phenomena that modern physics still regards as paradoxical emerge, including quantum phenomena, the special properties of Dark Matter, as well as the mechanism linking consciousness, the Soul, and the material world.

This perspective currently remains a philosophical–physical hypothesis. Its purpose is not to replace the standard models of modern physics, but to provide a unified theoretical framework in which spacetime, matter, Dark Matter, consciousness, and the process of informational transformation may be viewed as different levels of manifestation of the same deeper foundation.

According to the **TLD (Timeless Domain)** hypothesis, TLD is a foundational domain of reality that human beings are almost unable to perceive directly through the senses or through current methods of physical measurement. In contrast, the 4D world is the phenomenal domain that science has studied for centuries, discovering postulates, laws, and mathematical models describing the operation of matter, energy, space, and time.
Although very different in nature, TLD and the 4D world are not two separate realities, but two continuous domains of the same process of actualization. It is precisely this continuity that gives rise to an **Interface Domain**—the transitional region between the timeless reality of TLD and the spacetime reality of the physical world. This domain arises from the principle of uninterrupted continuity, necessarily connecting the two domains into a single reality.
The Interface Domain does not belong entirely to TLD, but neither is it yet entirely the 4D world. Here, characteristics of both domains coexist. On the one hand, structural information still retains its foundational nature and has not yet been fully materialized, as in TLD; on the other hand, the first manifestations of time, informational curvature, gravity, and matter have begun to appear. Therefore, the Interface Domain may be regarded as the place where the transformation from the invisible to the visible, from potentiality to actuality, and from pure information to physical phenomena takes place.
According to the TLDmodel proposed in this book, many difficult-to-explain phenomena of modern physics may originate precisely within this Interface Domain. This may be where **"thin time"** t_1, **proto-gravity, proto-dark matter (DM1)**, and **pre-geometric structures** appear before being fully actualized in 4D spacetime. Therefore, the Interface Domain serves not only as a bridge between the two domains of reality, but also as a key to understanding the process of material actualization and the transition from structural information to the phenomenal world.

In addition, dark matter is known to produce gravitational effects, although its nature has not yet been identified. In many theoretical physics models, dark matter has been hypothesized to be related to supersymmetric particles or other new particles beyond the Standard Model. However, the possibility cannot be excluded that dark matter is not composed of material particles in the ordinary sense, but instead reflects geometric or informational properties of a deeper intermediate domain.

From the perspective of the TLD model, this suggests that the **Interface Domain** may be an appropriate place for interpreting the origin of gravity and dark matter. At the same time, from a philosophical perspective, the **Five Aggregates**—which are important components constituting the mental life and karmic consciousness of each individual—appear to have several structural correspondences with the additional dimensions from D6 to D11 in M-theory, although these two concepts are not completely identical.

According to the hypothesis proposed in this book, this correspondence is not intended to identify Buddhism with M-theory, but merely to suggest that the additional dimensions D6–D11 may be regarded as a possible mathematical representation of the portion of the Interface Domain **approaching the four-dimensional world**. According to this hypothesis, many phenomena commonly attributed to additional dimensions—such as gravity, dark matter, and pre-material structures—may be understood as manifestations of the Interface Domain during the process by which primordial information is transformed into the material world.

A. SOUL OR KARMIC CONSCIOUSNESS (KARMIC CONSCIOUSNESS / SOUL)

If TLD and the 4D material world (or 4D as described within multidimensional theories such as 11D/M-theory) are indeed two domains of different natures that nevertheless interact with one another, then, according to the model proposed in this book, the existence of an **Interface Domain** is a logical consequence. This domain serves as a bridge, allowing information to be transformed and transmitted between the timeless informational domain and the material world of spacetime.

According to the hypothesis of the TLD model, the foundation of the Interface Domain is manifested through **Karmic Consciousness**, also called the **Soul**. Karmic Consciousness does not belong entirely to TLD, but neither does it yet belong entirely to the material world. It is an intermediate informational structure possessing characteristics of both domains, thereby enabling it to maintain the continuity of information between TLD and the brain.

In this model, Karmic Consciousness not only reflects the mental state of a present life but is also hypothesized to preserve the entire informational structure of each individual. This structure includes imprints formed during the present life as well as information accumulated from previous lives, if their existence is real. Therefore, Karmic Consciousness is regarded as a **continuous informational structure**, ensuring continuity in the process of learning, transformation, and development of the **proto-individual** throughout multiple stages of actualization.

From this perspective, the Soul is not a supernatural entity separated from the laws of nature, but an intermediate component of the process of informational transformation. It is through Karmic Consciousness that the informational structures of TLD may gradually be manifested in the material world, while the experiences and learning acquired during physical life are simultaneously fed back to the primordial informational foundation.

As presented in previous chapters, the TLD model proposes a new perspective on the foundation of spacetime. According to this model, the four-dimensional world (4D) is not primordial reality, but an **"educational environment,"** in which deviated informational structures are stratified, revealed, and transformed. The emergence of spacetime therefore carries not only physical significance but also a functional meaning for the entire process of informational evolution.

From this perspective, time is not merely a coordinate axis in physics. Time is the mechanism that allows three fundamental processes to occur:

1. **Classification and stratification of information**, so that different structures can be identified and separated.
2. **Revelation of deviated information**, allowing deviations that remain latent to become apparent through processes of motion and interaction.
3. **Correction and transformation of information**, including:
    - (a) correction directly within the TLD domain;
    - (b) actualization into the 4D world as an environment for comprehensive learning and transformation.

According to this hypothesis, the emergence of time may proceed through several levels.

At the initial stage (**proto-time,** t_0), time exists only as a latent possibility, sufficient to establish the ordering and stratification of information but not yet forming a physical flow of time.

At the next stage (**minimal time,** t_1), time becomes more clearly defined, allowing deviated information to become manifest and to persist in a semi-material state. At this stage, the amount of time is still very small, insufficient to form ordinary matter or the rich energetic processes found in the 4D world.

If this hypothesis is correct, one noteworthy possibility is that the combination of deviated information and minimal time may manifest as **Dark Matter**. From this perspective, Dark Matter is not merely an unknown physical component, but also an intermediate state between the timeless informational domain and the material world. It still possesses gravitational properties but interacts very weakly electromagnetically, consistent with many observed characteristics of dark matter.

This hypothesis also suggests another interpretation of why Dark Matter constitutes approximately 27% of the total mass–energy content of the universe. Rather than viewing it merely as a "missing" material component, the TLD model asks whether this large proportion might reflect the very process of informational transformation before ordinary matter is formed.

If so, the Interface Domain between TLD and spacetime may be the place where many phenomena that modern physics still regards as paradoxical emerge, including quantum phenomena, the special properties of Dark Matter, as well as the mechanism linking consciousness, the Soul, and the material world.

This perspective currently remains a philosophical–physical hypothesis. Its purpose is not to replace the standard models of modern physics, but to provide a unified theoretical framework in which spacetime, matter, Dark Matter, consciousness, and the process of informational transformation may be viewed as different levels of manifestation of the same deeper foundation.

Figure 9. A. Connection Between the Soul and the Brain Through Synaptic Potential

The Soul, if regarded as a subtle material structure, may be composed of elementary particles such as **Dark Matter and Neutrinos**, with additional **Electrons** attached to it, thereby giving it a negative electrical potential (−). Through this negative charge, the Soul may bind to the neural synaptic membrane, which carries a positive potential (+) in its resting state, creating a form of bioelectrical attraction between two entities belonging to different levels of existence—the physical brain and the subtle matter of the Soul.

Process of Separation When the Electrical Potential Changes

When a neuron is activated to transmit a signal, the electrical potential at the synaptic membrane becomes negative (−). At the same time, electrons on the surface of the Soul are also repelled because of the loss of electrostatic binding, causing the Soul's potential to become more negative, although this potential is very weak compared with the fluctuations of the biological electrical potential of the nervous system.

When the brain enters a state of complete loss of electrical potential—as in biological death—the synapses cease to function, and the electrostatic attraction between the Soul and the brain disappears. Electrons are gradually lost, allowing the Soul to naturally detach from the region of neural binding. This explains why the phenomenon of the **"Soul leaving the body"** occurs more easily after brain death than during life.

A prolonged state of reduced neural potential, known in neuroscience as **Long-Term Depression (LTD)**, may create similar electrical-potential conditions, leading to the possibility of partial separation of the Soul during deep meditation or a near-death experience.

Φ,ψ:wave functions\Phi,\psi: \text{wave functions} σ:total number\sigma: \text{total number} Q:superchargeQ: \text{supercharge} μ:indices\mu: \text{indices} ρ:energy density\rho: \text{energy density} p:pressure of CK<0p: \text{pressure of CK}<0 Λ:cosmological constant\Lambda: \text{cosmological constant} SUSY:supersymmetry

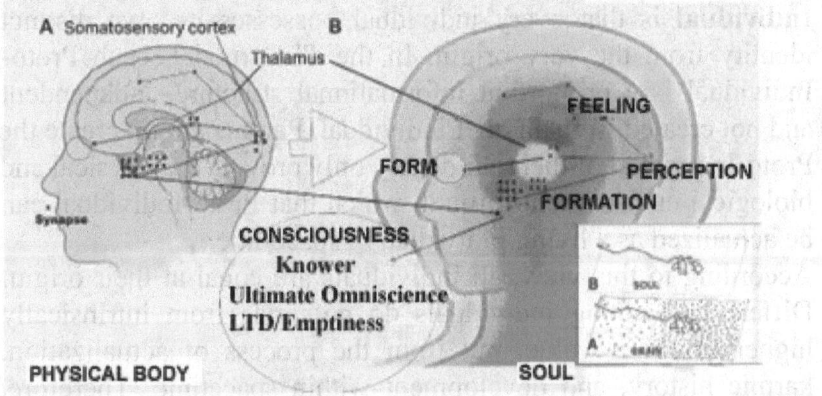

A: Neural pathways and neural connection networks in the sensory cortex (the dots represent sites of neural connections).

B: Neural pathways and neural connection networks imprinted on the Soul.

C. The Soul in its *primordial* state before recording memory or karmic consciousness. The white circular nodes represent **pure neutrinos** and **standardized neutrinos**, arranged according to their natural characteristics. In three-dimensional space, these neutrinos are densely arranged like the **structure of carbon atoms in a diamond crystal**: transparent and without forming an obstructive network, thereby neither impeding nor limiting the flow of information.

C. THE SIGNIFICANCE OF THE PROTO-INDIVIDUAL

According to the TLDmodel, individualization is not the ultimate purpose of the process of creation. It is only a necessary stage in the process of education and the restoration of integral information. When information remains in a pure state and has not yet become deviated, TLD has no need for stratification. However, when information begins to deviate from its primordial equilibrium state, the process of stratification is activated so that the information can be identified, actualized, and gradually self-corrected.

It is precisely at this stage that the **Proto-Individual** (primordial individual entity) appears. The Proto-Individual is not a complete soul or an individual in the ordinary sense. It is only a cluster of information that has become individualized within TLD after stratification begins. It is the earliest form of individuality, but it does not yet possess a body, a brain, biological memories, or an ego.

One of the important implications of the concept of the **Proto-Individual** is that every individual possesses its own distinct identity from the very origin. In the TLD model, each Proto-Individual is a primordial informational structure, independent and not created from another individual. Parents do not create the Proto-Individual of their child; they only provide the physical and biological environment through which that Proto-Individual can be actualized as a living individual in spacetime.

According to this view, all individuals are equal at their origin. Differences among individuals do not arise from intrinsically higher or lower value, but from the process of actualization, karmic history, and development within spacetime. Therefore, every individual possesses an intrinsic dignity and cannot be regarded as the product or property of any other individual.

This idea bears some similarity to Buddhist teachings. The Buddha taught that all sentient beings share a common true foundation, although they manifest as countless different individuals. According to this interpretation, all beings arise from the same ultimate foundation without thereby losing their individuality. Each individual is a distinct manifestation of that same foundation.

This view may also be compared with modern embryology. Parents carry and transmit genetic material and provide the biological conditions for embryonic development, but they do not actually "manufacture" the genetic material of the individual. The inherited genetic material has its origin in TLD. In the TLD model, the Proto-Individual is the primordial informational foundation that combines with the process of biological development to form a complete human being.

Actualization of Time

As the process of **temporalization** continues to develop, time is gradually actualized from **proto-time (t_0)** to **minimal time (t_1)**, corresponding to Dark Matter and the Soul, and then to **full time (t_2)**. The Proto-Individual enters the Interface Domain, where it begins to acquire new characteristics such as the capacity to preserve memories, developmental tendencies, and karmic consciousness. When combined with a body in the 4D world, the Proto-Individual manifests as a living individual.

The process can be summarized as follows:

TLD → Proto-Individual → Temporalization → Soul in the Interface Domain → Body and Brain → Living Individual in the 4D World.

In this model, the 4D individual is only the phenomenal manifestation of a deeper prototype. The individual possesses a body, brain, biological memories, and a particular social environment, whereas the Proto-Individual is the underlying informational structure that maintains the continuity of individuality throughout the entire process of actualization.

Because TLD and the 4D world are orthogonal to one another, a living individual cannot easily perceive their own prototype directly. Everyday consciousness is limited by the senses, neural activity, and linear time. Human beings therefore primarily perceive phenomena that have already manifested within the physical world.

Conversely, the Proto-Individual may influence the individual through the Interface Domain. This influence does not occur as a physical signal transmitted through space, but rather as a form of informational resonance between the foundational and phenomenal levels. Intuition, creative inspiration, innate tendencies, deep memories, or forms of awareness beyond ordinary reasoning may be regarded as manifestations of this resonance.

The formation of knowledge can also be viewed from this perspective.

According to Buddhist psychology, the cognitive process begins with **Form (Rūpa)**, that is, contact between the senses and the external world. These signals are received through the five senses and transmitted to the brain.

At the stage of **Feeling (Vedanā)**, the brain receives and processes the initial sensory signals.

At the stage of **Perception (Saṃjñā)**, information is rapidly compared with pre-existing reflexes and experiences. Many survival responses may occur almost instantaneously, without requiring conscious thought.

Next comes **Mental Formation (Saṃskāra).** The brain integrates multiple sources of information into stable patterns of neural activity. These patterns are encoded as electrical impulses and changes within synaptic networks and are subsequently stored in the brain's memory system. At this stage, however, **knowledge has not yet appeared**; only **neural information codes** have been formed.

According to the TLD hypothesis, a corresponding copy of these information codes is reflected onto the Soul within the Interface Domain. There, the Soul does not itself generate knowledge, but instead serves to transfer these information codes to the Proto-Individual.

It is the Proto-Individual that performs the process of interpreting information. New codes are compared with the entirety of the informational structures already present. If the new information corresponds to existing structures, it is recognized as familiar knowledge. If differences are present, the Proto-Individual begins a process of reorganization and integration. It is this process that gives rise to **learning**, **creativity**, and the **expansion of knowledge**.

According to this view, the brain does not directly create knowledge. The brain primarily encodes, stores, and processes information. Knowledge appears only when these information codes are interpreted and integrated by the Proto-Individual through the Interface Domain.

This model also explains why human beings may experience intuition, premonition, or forms of awareness that go beyond purely logical reasoning. Such phenomena do not necessarily negate the role of the brain; rather, they reflect cooperation among three levels of reality: **the brain in the 4D world, the Soul in the Interface Domain**, and **the Proto-Individual in TLD**.

According to this hypothesis, therefore, human cognition is neither merely the activity of neurons nor merely the independent activity of the Soul. Rather, it results from continuous coordination among three levels of the same process of informational actualization. It is precisely this coordination that transforms information into knowledge, and knowledge into the lived experience of each individual.

In summary, according to the TLDmodel, throughout the **temporalization** of structural information into a living individual in the physical world, the **Proto-Individual is not transformed into another entity**. It continues to exist as the underlying informational structure within TLD. What becomes actualized through the Interface Domain and the 4D world is only the **temporalized version** of the Proto-Individual.

In other words, the Proto-Individual is the **prototype**, whereas the individual in the 4D world is the **phenotypic manifestation** of that prototype under the influence of time, matter, and biological conditions.

The Proto-Individual continuously maintains a connection with the individual in the 4D world through the Interface Domain. However, this relationship is not completely symmetrical. From the TLD side, the Proto-Individual may continuously influence the individual through intuition, innate tendencies, inspiration, inner awareness, and processes involved in the formation of knowledge. Conversely, from the perspective of the 4D individual, access back to the Proto-Individual is restricted by the orthogonality between TLD and spacetime, as well as by the limitations imposed by the senses, the brain, and linear time.

This relationship can be illustrated through the analogy of a diplomatic mission within a country. The head of state who sends the diplomatic delegation represents the Proto-Individual in TLD; the diplomatic mission within the host country represents the individual in the 4D world. The mission always depends upon the head of state. Every movement of the mission is determined by the head of state. However, the mission itself cannot act backward to alter the head of state. Only when the hierarchical relationship changes can the head of state who established the diplomatic mission be recognized.

According to the TLDmodel, the relationship between the Proto-Individual and the 4D individual is similar. The individual in the phenomenal world is the manifestation of a deeper informational structure. The existence and development of the individual are always connected with the Proto-Individual, while the ability to perceive the Proto-Individual directly is limited by the very conditions of spacetime.

Therefore, the path of spiritual practice, the development of wisdom, and the purification of information are not intended to create a new entity, but to help transform the phenomenal individual; this transformation is always perceived by the structure of the prototype within TLD. As informational deviation is gradually transformed, the distance between the phenomenal individual and the Proto-Individual is progressively reduced, bringing it closer to the integral structure of TLD.

Within the **Proto-Individual** model, the process of spiritual cultivation, learning, and transformation of the Soul during a lifetime does not alter TLD as an absolute reality; rather, it **adjusts the informational structure of the Proto-Individual itself**. In the language of the model, the degree of **"informational curvature"** is gradually reduced when an individual lives ethically and develops wisdom and compassion; conversely, it increases when the individual continues to generate karma and deviate from the state of integrity.

If this is so, then after death, what determines the next state is not an **external judgment**, but rather the **informational state of the Proto-Individual itself**. The system requires no judge to impose a sentence; the individual's informational structure itself determines its next equilibrium state.

According to this view, the concept of an **intermediate period after death** found in many religious traditions may be understood as **a transitional stage** during which informational reorganization and reconnection take place before entry into a new state of existence. "Judgment," if it exists, is not a courtroom proceeding in the legal sense, but rather **the objective manifestation of the karmic force and informational structure that the individual has created**.

This is also close to many Buddhist ideas: **karma itself leads to its fruit**, rather than an external being deciding reward or punishment. It should be emphasized, however, that this is **a philosophical interpretation within the TLD model**, not a scientifically established conclusion or the only possible interpretation of religious traditions.

D. OTHER ROLES OF THE SOUL

a) Memory and Memory Retrieval

According to the model proposed in this book, one of the important roles of the **Soul (Karmic Consciousness)** is to participate in the formation, preservation, and retrieval of memory.

The model hypothesizes that, during the acquisition of knowledge, the Soul directly stores the complete content of each experience and continuously records a **structural map** of the neural pathways formed within the brain. This map reflects the process of information processing from the sensory organs to the cerebral cortex through stages corresponding to the Five Aggregates: **Form, Feeling, Perception, Mental Formation, and Consciousness**.

Within this process, **Form** reflects initial sensory information; **Feeling** records the emotional value of an experience; **Perception** participates in unconditioned reflex recognition; **Mental Formation** contributes to the formation of responses, habits, and processing tendencies; and **Consciousness** integrates the entire process into conscious experience. According to this model, the role of **Perception** may in many cases be only temporary, primarily producing associations and immediate emotional responses before information becomes stabilized within memory networks.

When a present event shares characteristics with a past experience, these signals may activate memory traces (**memory engrams**) stored within the **hippocampus**, whose role is to **index and support memory retrieval**.

According to the hypothesis of the model, simultaneously with this process, the Soul uses the structural map previously recorded to reactivate the corresponding neural pathways. Through coordination between brain activity and the informational map within Karmic Consciousness, an old memory can be reconstructed and returned to consciousness.

According to this view, the Soul does not store each memory as if it were a data warehouse. Rather, it preserves the **organizational structure of the neural pathways** that produced that memory. This structure functions as an index or positioning map, enabling the nervous system to locate the appropriate synaptic network among countless possible connections. Without such a guiding mechanism, accurately retrieving one specific memory from billions of neurons and trillions of synapses would be extraordinarily difficult.

Within the TLD model, this hypothesis is extended further by proposing that this structural map exists not only within the brain during one lifetime, but is also maintained within the **Proto-Individual** as part of Karmic Consciousness. If correct, this would provide a theoretical mechanism for explaining continuity of learning and experience beyond the limits of a single lifetime.

b) Certain Unusual Phenomena of Consciousness May Be Viewed from a Different Perspective

For example, in cases of **Dissociative Identity Disorder**, the TLD model proposes an interpretation different from current neurological and psychiatric models. According to this hypothesis, a significant change may have occurred in the coordinating mechanism of the Soul (Karmic Consciousness) over the neural network. In such a situation, the informational structure currently serving as the individual's dominant coordinating structure may temporarily lose influence, allowing other informational structures to manifest more clearly.

The model further extends the hypothesis that a developing nervous system may, in rare circumstances, interact with more than one **Proto-Individual** or with multiple distinct informational structures. Under this possibility, alternation or competition in the coordination of neural activity could produce different states of consciousness within the same individual, manifesting as multiple-personality pathology.

This hypothesis is suggested by the fact that biology has documented certain abnormalities during fertilization involving more than one sperm in embryonic development, suggesting that the formation of an individual does not always follow a simple pattern. Such phenomena suggest the possible existence of multiple Proto-Individuals or multiple Souls within the same body. Within the TLD model, they are used only as a basis for asking whether the relationship between the embryo and informational structures may be more complex than ordinarily assumed.

Consequently, phenomena such as dissociative identity disorder, sleepwalking, or certain unusual states of consciousness may be investigated from a new perspective. According to this view, such phenomena need not reflect only the disturbance of a single informational structure, but could involve changes in the coordinating roles of multiple informational structures interacting with the same nervous system. This remains a theoretical hypothesis of the TLD model and currently lacks experimental evidence confirming it.

Out-of-Body Experience

The phenomenon of the **Out-of-Body Experience (OBE)** has previously been discussed by the author in an earlier published book. Within the TLD model, this phenomenon is regarded as a possible consequence if the Soul (Karmic Consciousness) truly functions as an intermediary component between the informational domain and the brain.

According to the hypothesis of the model, under ordinary conditions the Soul is closely linked to the nervous system through the brain's synaptic network. This connection is hypothesized to occur through an extremely subtle physical mechanism, possibly involving interactions between electrical charges at synapses and subtle-material components of the Soul. In previous work, the author hypothesized that this mechanism could involve coupling between electrons at synapses and a foundational component of the Soul, such as a **sterile neutrino** or another presently unknown fundamental particle. This is only a hypothesis intended to describe a possible interaction mechanism between the two domains of reality and currently has no experimental confirmation.

When the connection between the Soul and the synaptic network is weakened or temporarily interrupted—for example, during certain near-death states, very deep meditation, under the influence of certain psychoactive substances, or spontaneously—the Soul may become detached from its direct binding to the physical body. According to the TLD model, the entire structure of Karmic Consciousness, including components corresponding to the Five Aggregates, remains preserved as an informational whole even though it is no longer directly attached to nervous-system activity.

Because the Soul is hypothesized to belong to the Interface Domain and is not completely constrained by the spatial limitations of the 4D world, the model proposes that it may exhibit **non-local characteristics**. At the same time, the Soul continues to maintain its normal relationship with the Proto-Individual in TLD. This characteristic is used to explain why many people reporting OBE describe observing their own bodies from outside or simultaneously perceiving multiple locations in space.

However, according to this hypothesis, although the Soul continues to preserve awareness and memory, once detached from the nervous system it can no longer interact mechanically and directly with matter. Therefore, the Soul cannot generate significant mechanical force to move objects, produce sounds, or perform physical movements like a biological body. Any interaction with the physical world, if it occurs, must take place through re-establishing a connection with the nervous system or through mechanisms that science does not yet understand.

This model is proposed as a theoretical framework for interpreting reports of out-of-body experiences in a manner consistent with the TLD hypothesis.

E. INTERPRETING THE FIVE AGGREGATES THROUGH STRATIFIED QUANTUM MECHANICS: THE FIVE AGGREGATES, HIGHER-DIMENSIONAL SPACE, AND THE INTERFACE DOMAIN

In String Theory and M-theory, the four familiar dimensions of spacetime—three spatial dimensions and one temporal dimension—are the dimensions in which the fields and particles of the Standard Model manifest directly and can therefore be experimentally investigated. In contrast, the additional dimensions from D5 to D11 are hypothesized to exist as compactified dimensions or brane structures at extremely small scales and therefore cannot yet be directly observed. Their presence serves not only to maintain the mathematical consistency of the theory, but also to eliminate quantum anomalies and unify the fundamental interactions. They also lead to predictions of many new particles, particularly supersymmetric partners, although no experimental evidence has yet confirmed their existence. According to the TLD model proposed in this book, the additional dimensions D6–D11 of M-theory nevertheless remain within the physical domain because they are described through geometry, fields, and energy.

1. D6 (Form Aggregate): The Imperceptible Energy Scaffold and Neural Synaptic Matrix

In the IPTM hypothesis, the **Form Aggregate resides in the sixth dimension (D6)**—the first hidden spatial dimension immediately adjacent to D5, the dimension associated with gravity—and functions as an **Imperceptible Energy Scaffold**.

The Form Aggregate is not biological matter, but a quantum structural network woven from a **Dark Substrate** (Dark Matter/Dark Energy) combined with a stream of **neutrinos**. Because neutrinos have extremely small mass, move at nearly the speed of light, and interact only very weakly with ordinary matter, this Form-Aggregate network can penetrate physical barriers and deeply envelop and integrate with the organism's entire nervous system.

The connection between the invisible D6 structure and the 4D physical body is established at **synapses** and **peripheral neuronal terminal endings**. From the cerebral cortex and thalamus to nerve endings in the skin, skeletal muscles, and viscera, the neutrino network of the Form Aggregate continuously undergoes weak interactions with streams of electrons moving through ion channels. This D6 quantum network functions as a macroscopic geometric template that guides cellular development and forms an invisible outline of the organism's body.

2. D7 (Feeling Aggregate): Probability-State Filter and Primary Sensory Determinant Calculation

Immediately following the energetic envelope of the Form Aggregate is the **Feeling Aggregate residing in the seventh dimension (D7)**. The neuroanatomical process—transmission from the sensory organs through the **thalamus** to primary sensory cortical regions such as **V1, S1, A1, etc.**—is the 4D manifestation of the processing of the raw-sensory-data density matrix occurring at D7.

When the senses receive stimulation, the synaptic matrix at D6 generates a quantum superposition containing numerous complex fluctuating amplitudes. To transform this chaotic stream of electrical impulses into a meaningful data state, D7 receives and arranges them into a **Feeling-Aggregate Density Matrix, ρ_{D7} }**.

Here, XX represents neither pleasure nor suffering.

The off-diagonal components cij are complex quantum interference phases representing disturbance. At this stage, to package the data and transmit it to higher levels, D7 activates a **State-Freezing** algorithm by calculating the **determinant**:

$\det_{\rho D7}$. The determinant calculation compresses the entire multidimensional probabilistic vector space of neural electrical impulses into a single real scalar value. This process completely eliminates the interfering phases,

$cij \rightarrow 0$,

freezing the complex probability matrix into a minimally compressed raw result ready to be transferred to **D8 (Perception Aggregate)** to activate automatic survival reflexes.

3. D8 (Perception Aggregate): Quantum Snapshot Algorithm and Unconditioned Reflex Mechanism

The **Perception Aggregate residing in the eighth spatial dimension (D8)** functions as an **automatic quantum graphical storage system**, responsible for operating the unconditioned-reflex algorithm whose ultimate purpose is to protect the integrity of the biological body. At this primary stage, the Perception Aggregate is not yet knowledge or a linguistic concept.

Immediately after the raw sensory matrix is compressed through the determinant calculation at D7, D8 applies a **Projection Operator P_{D8}** to execute a **Quantum Snapshot**:

$$PD8 = |\Phi Pattern\rangle\langle\Phi Pattern|$$

This projection operator scans the real scalar value received from D7 and immediately compares it with a stored matrix system of **Pre-programmed Survival Schemas**.

For example, when a foreign object flies directly toward the eye at high speed, the neural electrical impulse transmitted from the retina through V1 is projected directly toward D8. D8 recognizes that this impulse pattern corresponds to a threat pattern, and the projection operator P_{D8} activates the unconditioned reflex of closing the eyes, sending the command immediately back down to the musculoskeletal matrix at D6 without waiting for logical analysis by the brain.

The result of this Quantum Snapshot command is compressed into a flat **holographic interference pattern** and transferred to **D9 (Mental Formation Aggregate).**

4. D9 (Mental Formation Aggregate): Nonlinear Integration Operator and the Birth of the Ultimate Information Code (UIC)

The **Mental Formation Aggregate residing in the ninth spatial dimension (D9)** is the great **State Synthesizer**, where a **Nonlinear Integration Operator I_{Int}** operates to consolidate all discrete variables into a single unified informational entity.

The integration algorithm performs a multilayer tensor product combining three fundamental data sources as input parameters.

Here, the environmental matrix represents the physical environmental constants fed back from 4D/5D space, including pressure, temperature, and the mechanical state of the body. The nonlinear operator I_{Int} compresses the entire matrix of biological data, automatic reflexes, and environmental context into a single quantum impulse sequence called the **Ultimate Information Code (UIC)**.

The UIC is structured as an entangled multistate qubit string containing the integrated information of:

|Form,Feeling,Perception,Mental Formation,Consciousness

This UIC string carries the entire structural information of the organism and is ready to be transferred to the **D10 operating system (Consciousness Aggregate).**

5. D10 (Consciousness Aggregate): Cognitive Operating System and Gateway of Quantum Decoding

The **Consciousness Aggregate residing in the tenth spatial dimension (D10)** operates as a **Master Cognitive Operating System**. It functions as a **Data Auditor** positioned between the dynamically changing 4D world and the eternal informational domain of TLD.

The central task of D10 is to receive the **Ultimate Information Code qubit string** transmitted upward from the Mental Formation Aggregate (D9). To transform this raw qubit string into knowledge and generate volitional responses, D10 performs a structural transformation operator, continuously making differential comparisons between the current |UIC⟩|UIC\rangle string and previous state traces.

When structural correspondence is detected, D10 directs and sends the information string through the orthogonal axis of the Wick rotation, transmitting it directly to **D11 (The Knower)**. The Consciousness Aggregate is therefore the final gateway directing physical information toward the highest ontological level.

6. D11 (The Knower / Soul): The Highest Control Axis and the 4D → TLD → 4D Loop

The **eleventh dimension (D11)—The Knower, or the Soul—is the highest control axis of the Interface**, directly adjoining the TLD domain. D11 functions as a **Supreme Directional Vector Operator.**

a) Outgoing Direction: 4D → TLD

When an event begins to occur within the neural synaptic matrix system at D6, its code sequence passes through the structure of the Aggregates and reaches Consciousness (D10). D10 pushes this data through the Interface Firewall filter to D11. There, through the orthogonal bridge of the Wick rotation, **The Knower (D11)** instantaneously synchronizes this real-world data sequence with the original TLD server.

b) Core Processing and Selective Filter

TLD receives the actual data from D11 and operates a **selective-filter algorithm** to determine the optimal eigenvalue consistent with the global law. The 4D universe cannot independently access the total data contained within TLD unless such access is permitted.

c) Return Direction: TLD → 4D

The optimal solution is reflected back from TLD through D11 in the form of a **Volitional Pressure**. D11 receives this pressure and applies a **unitary matrix operator** to the entire Five-Aggregate system beneath it:

$|\psi\rangle_{new}=U(\theta).|UIC\rangle$ This operator rotates the quantum phases and immediately sends a control command directly down to D6, completely eliminating the quantum interference phases:

$c_{ij}\rightarrow 0$.c_{ij}\rightarrow0.

The probabilistic wave function existing in superposition within the physical 4D universe is thereby forced to collapse instantaneously into a single definite reality, giving rise to knowledge and a volitional behavioral response.

Ontological Clarification Concerning D11

D11 remains part of the material aspect of the Interface and is not yet the ultimate source-domain of TLD.

Although D11 (The Knower / Soul) resides at the highest level of the Five-Aggregate structure, it still operates through the mathematical structure of a directional vector operator and remains constrained by eleven-dimensional space (M-theory). This means that D11 remains within the domain of **Duality**, where a distinction still exists between the **"knowing subject" (The Knower)** and the **"object being known" (the UIC code transmitted upward)**. It operates as a supreme processor composed of Dark Matter or neutrinos; therefore, it remains an intermediary instrumental entity.

According to the hypothesis proposed in this book, there may be a **structural correspondence** between the Buddhist concept of the Five Aggregates and the additional dimensions of M-theory, although the two systems are constructed upon entirely different foundations.

In Buddhism, **aggregate (skandha)** means accumulation, aggregation, or covering. The Five Aggregates are not ultimate reality but conditioned structures that continuously change while simultaneously obscuring the true nature of reality. Therefore, the Five Aggregates are not Suchness itself, but rather the manifest level through which human beings perceive the world and themselves.

In the TLD model, a similar role is assigned to the **Interface Domain**. It is the intermediary domain where primordial structural information begins to undergo temporalization, stratification, and transformation into physical as well as psychological phenomena. All manifestations of matter, energy, gravity, and consciousness first emerge within this domain before becoming fully actualized in four-dimensional spacetime.

From this perspective, the Five Aggregates may be regarded as the **manifestation of the Interface Domain at the existential level of human beings**. What Buddhism calls attachment to the Five Aggregates may, in the language of the TLD model, be interpreted as identification with temporalized and stratified informational structures rather than recognition of the integral informational foundation underlying them.

If this hypothesis is accepted, then the additional dimensions from five through eleven in M-theory may be regarded as **a possible mathematical representation** of the Interface Domain rather than of TLD itself. This also suggests another possible understanding of gravity and Dark Matter. Rather than being phenomena belonging exclusively to the physical world, they may reflect properties of the Interface Domain, where primordial information begins to manifest as geometric structures and ultimately becomes actualized as matter.

According to this view, **TLD does not correspond to the eleven-dimensional space of M-theory**. Rather, if a correspondence exists, it is the **Interface Domain** that is the closer candidate for correspondence with the additional dimensions of String Theory, because it is there that pre-geometric structures begin to acquire characteristics describable in geometric and physical language.

Therefore, this may provisionally be represented in concise form as:

TLD →Interface (DM/five skandha /6-11D)→ 4D

Conclusion
The correspondence is absolute:
Information in TLD gives rise to Geometry → Geometry manifests as Vibrating Strings → Strings give rise to Matter.
According to String Theory and M-theory, the mathematical consistency of the equations requires space to possess more than the familiar four dimensions. In the prevailing models, the four dimensions of spacetime are where the fields and particles of the Standard Model manifest and can be experimentally investigated. By contrast, the additional dimensions appear as geometric structures at extremely small scales, serving to maintain theoretical consistency and permitting the unification of fundamental interactions.
However, these additional dimensions have not yet been associated with any specific physical or biological entity that can be directly tested. They are primarily regarded as components of mathematical models rather than experimentally confirmed levels of existence.
The TLD model proposed in this book offers a different interpretation. Instead of regarding the additional dimensions merely as abstract geometric dimensions, the model hypothesizes that they may correspond to the **Interface Domain**—the intermediary level connecting the primordial informational domain (TLD) with the physical 4D world.
From this perspective, the **Soul (Karmic Consciousness)** is the informational structure operating within the Interface Domain. The Soul does not belong entirely to TLD because it already carries the imprint of time and experience, but neither does it belong entirely to the physical world because it continues to maintain non-local characteristics and the capacity to preserve information throughout the process of actualization.
According to this hypothesis, each additional dimension of M-theory may correspond to a functional level of Karmic Consciousness. Accordingly, the informational layers corresponding to **D6 ↔ Form, D7 ↔ Feeling, D8 ↔ Perception, D9 ↔ Mental Formation, and D10 ↔ Consciousness** may be

regarded as five operational levels of the Interface Domain, while the highest level, **The Knower**, directly adjoins TLD.
This interpretation does not alter the mathematical structure of M-theory; rather, it proposes a new physical and philosophical meaning for its additional dimensions. These dimensions would no longer be merely anonymous geometric dimensions, but would become levels of informational transformation through which the primordial structure of TLD is progressively actualized into consciousness, gravity, and ultimately matter.
According to the TLD model, the entire process may therefore be summarized as

TLD (canonical inồmation)
→ Interface **(Soul– Five Skandhá– 6-11D)**
→ **4D (Fields, påticles phenomena).**

With this approach, the TLD model is not intended to replace M-theory. Rather, it serves as an additional interpretive layer, suggesting a possible physical and ontological meaning for the additional dimensions predicted by mathematics but for which no unified interpretation has yet been established.

CHAPTER VII: INTERFACE DOMAIN (2): PARADOXICAL PHENOMENA OF QUANTUM MECHANICS

Summary

Quantum mechanics has achieved tremendous success in predicting microscopic phenomena and has provided the foundation for many modern technologies. However, despite its extremely high experimental accuracy, the theory still faces many difficulties in explaining the underlying nature of quantum phenomena. Issues such as quantum superposition, wave-function collapse, the EPR paradox, and the role of the observer remain fundamental questions. For this reason, Albert Einstein once maintained that quantum mechanics, despite its success, was still not a complete description of reality.

According to the TLD model proposed in this book, the Timeless Domain may serve as a deeper foundation underlying quantum phenomena. Within the framework of the ***Information Pre-Geometrization and Temporalization into Matter Theory (IPTM)******, quantum phenomena are regarded as intermediate states in the process through which structural information from TLD is transformed into matter within spacetime.*

In this process, the Interface Domain is where information exists in a transitional state and may simultaneously manifest in two aspects: ***informational structure*** *and* ***material particle******. Because it simultaneously carries characteristics of both TLD and the 4D world, the Interface Domain gives rise to phenomena that may appear paradoxical when viewed solely from the perspective of classical physics.*

From the informational perspective, the quantum state remains influenced by characteristics of TLD, such as non-locality, timelessness, and the absence of complete constraints imposed by distance or physical barriers. Conversely, when manifested as a material particle in 4D spacetime, the quantum state follows ordinary physical laws, including limits on propagation velocity and the interaction conditions of the material world.

According to this hypothesis, the wave–particle duality of quantum phenomena may be understood as a consequence of the quantum simultaneously possessing two aspects: one as an informational structure within the Interface Domain and the other as a material particle within spacetime. For this reason, a quantum entity may manifest as a wave while its information remains distributed throughout the Interface Domain, but manifest as a particle once the process of material actualization has been completed. Under this interpretation, phenomena such as quantum tunneling or quantum superposition do not necessarily represent violations of physical laws, but may instead be manifestations of a transitional stage between two domains of reality.

Another important feature of the TLD model is the asymmetric exchange of information between the Interface Domain and TLD. According to this hypothesis, information from the physical world is continuously updated to TLD. Conversely, information from TLD is transmitted back to the phenomenal world only selectively, depending on the informational structure and the state of the system under consideration.

Therefore, in experiments such as Young's double-slit experiment or the EPR paradox, when the state of a quantum particle changes as a result of measurement, information concerning that change is first registered within TLD. From this common informational foundation, the states of related quantum systems can be updated almost instantaneously within the Interface Domain before manifesting as observable outcomes in the 4D world. According to the proposed model, this provides a possible interpretive framework for explaining quantum non-locality without requiring the assumption that a physical signal propagates faster than the speed of light through spacetime.

In **Chapter V – Part I**, the book presented the hypothesis of **Karmic Consciousness (or Soul)** as an intermediate entity of the **Interface Domain**, serving as a bridge between the **Timeless Domain (TLD)** and the material world of spacetime (4D). According to the proposed model, the Interface Domain is precisely where primordial information begins to undergo temporalization, stratification, and transformation into physical phenomena.

In **Part II**, this model will be used to approach one of the most successful yet also most mysterious fields of modern physics: **quantum mechanics**.

It cannot be denied that quantum mechanics has opened a new era for science and technology. From transistors, lasers, and semiconductors to quantum computers, almost the entire modern technological civilization has been built upon quantum principles. However, alongside these enormous achievements, quantum mechanics still contains many fundamental questions for which no unified explanation has yet been found. Phenomena such as quantum superposition, wave-function collapse, the role of the observer, quantum tunneling, and quantum entanglement have remained subjects of debate for more than a century.

From this perspective, Albert Einstein's well-known view that *"the interpretation of quantum mechanics is still incomplete"* may not have been entirely wrong. What Einstein had not found may not have been an error in the quantum equations, but rather a deeper foundational layer that those equations reflect.

According to the view proposed in this book, that foundation does not belong exclusively to physics or mathematics, but is first and foremost an **ontological** question. To understand a phenomenon fully, it is necessary to distinguish between the **manifest world** and the **generative foundation** of that phenomenon. The manifest world is the 4D spacetime investigated by physics; the deeper foundation, according to the TLD model, is the timeless domain of primordial information, where matter, energy, geometry, and motion do not yet exist.

However, there is no absolute separation between these two domains. They are connected by the **Interface Domain**, where primordial information begins to acquire the characteristics of geometry, time, and physics before manifesting as the phenomenal world. Therefore, according to the hypothesis of the TLD model, quantum phenomena do not directly reflect TLD itself, but rather **processes occurring within the Interface Domain**, where information is transitioning from a timeless state into the material world.

If quantum mechanics truly marks the present boundary of physics, then that very boundary may indicate that its objects no longer belong entirely to the 4D material world. Characteristics such as non-locality, superposition, entanglement, and the role of measurement may be signs that quantum mechanics is approaching the boundary between the physical world and the Interface Domain. From this perspective, quantum mechanics is not merely the physics of extremely small particles, but may also be a gateway through which the existence of a deeper level of reality behind spacetime can be recognized.

I. POSSIBILITY OF TRANSFORMATION WITHIN THE INTERFACE DOMAIN

Information Asymmetry Principle

According to the TLD model proposed in this book, the **Interface Domain** is not only a bridge between the Timeless Domain (TLD) and the 4D material world, but also a **domain of information transformation** operating in both directions.

In the first direction, primordial structural information in TLD is gradually transformed into geometry, fields, forces, and ultimately matter within spacetime. This process has been described in the **IPTM model (Information Pre-geometrization and Progressive Conversion into Matter)**.

In the opposite direction, information arising from the phenomenal world is also continuously transmitted back to TLD through the Interface Domain. However, this process is not simply a transmission of data; it is simultaneously a process of **transformation**. The experiences, actions, and knowledge of an individual are encoded into informational structures, whereas the reverse direction involves decoding those informational structures so that they can be manifested as cognition, intention, and action in the 4D world.

According to the hypothesis of the model, this two-way transformation does not occur completely freely, but follows several fundamental principles.

(i) Outgoing direction: from 4D to TLD

Information arising in the phenomenal world is hypothesized to be continuously transmitted backward through the Interface Domain into TLD. From this perspective, almost every experience of an individual is recorded in the form of informational structures.

For each individual, this information is accumulated within the **proto-individual (Karmic Consciousness)** as traces of the process of learning and transformation. For the natural environment, the information is regarded as contributing to the general informational structure of the system.

According to this hypothesis, TLD continuously maintains an updated state corresponding to changes occurring within the phenomenal world.

(ii) Returning direction: from TLD to 4D

Conversely, the transmission of information from TLD down into the 4D world is hypothesized to be subject to very strict selection. Only information necessary for the processes of decoding, cognition, and actualization is permitted to manifest within the brain and the material world.

Information that is not directly relevant to the individual's process of actualization will not manifest in ordinary consciousness. Therefore, human beings perceive only a very small portion of the informational structures existing within TLD.

From this perspective, the 4D world does not directly perceive everything occurring within TLD, whereas TLD continues to register the entire process of the phenomenal world.

(iii) Exceptional cases

The model also does not exclude the possibility that, in certain very rare cases, some information from TLD may manifest more directly within the consciousness of certain individuals. Throughout history and in many cultural traditions, such individuals have often been described as prophets, sages, or persons possessing exceptional intuition.

According to the TLD model, this is only a hypothesis proposed to explain certain phenomena recorded in history and spirituality; there is currently no experimental evidence confirming it.

(iv) The state of enlightenment

In the TLD model, the state of enlightenment is understood as the gradual reduction of layers of obscuring information (**ignorance**), thereby allowing consciousness to gain more direct access to the Interface Domain.

This does not mean that the individual becomes completely identical with TLD, but rather that the degree of manifestation of canonical information progressively increases, while biased informational structures are progressively purified.

II. INFORMATION AND PARTICLES WITHIN THE INTERFACE DOMAIN

According to the proposed model, information within the Interface Domain can exist in two different modes of manifestation.

In a state closer to TLD, information has not yet been actualized into matter. In this state, spacetime in the physical sense does not yet exist; therefore, informational structures may manifest characteristics such as **non-locality** and independence from direct constraints of spatial distance and time.

As the process of actualization continues, information gradually transforms into quantum states that can manifest as fields and particles. From that point onward, these structures obey the laws of quantum physics and ultimately the laws of the 4D world.

From this perspective, the quantum particle is not the starting point of reality, but rather one stage in a continuous process of transformation from **information → geometry → field → particle → matter**. The Interface Domain is precisely where the transition between these two modes of existence takes place.

With the principles above, almost all paradoxes in physics become events that, from this perspective, can be regarded as determinate rather than uncertain or probabilistic.

According to this view, phenomena such as quantum interference, superposition, quantum entanglement, conservation of information, as well as the processes of birth and death, are no longer isolated phenomena. They can all be regarded as different manifestations of the same deeper principle: **the integral reflection between the timeless informational foundation (TLD) and the phenomenal 4D world**.

This is a philosophical–physical hypothesis of the TLD model. Its value will depend on the possibility of constructing a mathematical model for the Interface Domain and on its ability to generate predictions that can be experimentally tested in the future.

If this hypothesis is correct, then all quantum phenomena, from interference and superposition to entanglement, are no longer separate phenomena, but different manifestations of the same principle: **informational consistency between TLD and the 4D world**. TLD does not replace the existing laws of quantum mechanics, but is proposed as a deeper informational foundation from which the laws of quantum mechanics emerge as manifestations within spacetime.

This is a philosophical–physical hypothesis. Its value does not lie in replacing quantum mechanics, but in its potential to provide a unified framework for explaining why phenomena such as interference, entanglement, and the actualization of quantum states may all be regarded as manifestations of a common informational foundation extending beyond the conventional description of spacetime.

The popular saying "people act, Heaven watches" does not merely carry a simple moral meaning, but also contains a profound cosmological intuition. Human beings may conceal their actions from society, from the law, and even from their own conscience; but within the TLD model, no action disappears without leaving a trace. Every change in the 4D world has a corresponding informational reflection within TLD. Therefore, "Heaven watches" does not necessarily mean a personified observer, but rather the complete registration performed by the informational foundation. What human beings do within time is reflected within the timeless domain. This is the deeper foundation of karma: actions do not disappear, but become informational structures that continue to influence the process of existence.

III. PARADOXICAL QUANTUM PHENOMENA

Unlike relativity, which generally does not reach the threshold of TLD and whose phenomena, although often beyond intuition, can still be imagined through ordinary experience, quantum mechanics contains many phenomena that appear completely paradoxical to conventional concepts of space, time, and matter. Niels Bohr himself once emphasized that anyone who thinks he has completely understood quantum mechanics has perhaps not truly understood it. Meanwhile, Albert Einstein referred to quantum entanglement as "spooky action at a distance."

a) One of the most famous phenomena is **quantum entanglement**, or the EPR paradox. Two particles that once belonged to the same quantum system can continue to exhibit nearly instantaneous correlations even when they are separated by an enormous distance. According to the intuition of the 4D world, this appears to require a signal traveling faster than light, contrary to the limit imposed by relativity: two quantum particles are far apart, but both transmit information to TLD.

b) The second phenomenon is **quantum superposition**. A quantum particle can be described as existing simultaneously in multiple states before a measurement is performed: information and particle coexist within the Interface Domain.

c) The observer: Young's double-slit experiment shows that photons and electrons exhibit both particle-like and wave-like properties. Schrödinger's cat thought experiment pushes this idea to its extreme by describing a quantum system as being both alive and dead until its state is determined by measurement: phenomena occurring in 4D are naturally always observed from TLD.

d) The third phenomenon is **quantum tunneling**. A particle can appear on the other side of a potential barrier that, according to classical physics, it does not possess sufficient energy to overcome. This phenomenon has been experimentally verified and plays an important role in many processes of modern physics.

From the perspective of classical physics, these phenomena appear highly paradoxical. Even when extended into multidimensional models of modern physics, such as higher-dimensional spaces in string theory, their underlying nature still does not become truly intuitive.

e) THE OBSERVER PHENOMENON

According to the interpretation of the HTTM model, before a quantum state is actualized into a definite result within the 4D world, its informational structure is not yet completely bound to a single history of physical time t_2. This state can be represented by the principle of superposition:

|Ψ⟩=∑ici|i⟩|\Psi\rangle=\sum_i c_i|i\rangle

where cic_i is the probability amplitude of the state.

In the HTTM interpretation, the superposition state does not necessarily mean that a physical particle is simultaneously present at multiple locations in the classical sense. Instead, it represents an **informational structure that has not yet been actualized into a single 4D outcome**. Because it has not yet been attached to a definite t_2 history, the structure may exhibit wave-like behavior, non-locality, and the capacity for interference.

When a measurement occurs, or when an interaction is sufficiently strong for information about the system to be registered within the 4D environment, the system becomes associated with a definite outcome. In the language of HTTM, this process can be described as **complete temporalization into** t_2

The particle within the Interface Domain is continuously monitored, and its information is continuously updated to the server within TLD. Therefore, when it is KNOWN that there is an observer—or, more precisely, when time t_2 is added—the structural information immediately becomes a particle.

Thus, what is commonly called the "role of the observer" does not necessarily mean that the particle **knows** that a human being is looking at it. The decisive point is that the system has participated in an interaction through which information about its state is registered and becomes part of the physical history of the 4D world.

According to the HTTM hypothesis, this can be interpreted more deeply by proposing that **measurement itself is a special case of the temporalization of information**: while the structure has not yet been completely locked into t_2, it retains a spectrum of possibilities; once the information becomes attached to t_2, one possibility becomes a definite physical event.

Therefore:

From this perspective, **wave behavior** reflects the domain of informational possibilities that have not yet been completely localized within a definite 4D history, whereas **particle behavior** reflects a state that has been actualized into an event possessing position, energy, and before–after relations within t_2.

The phenomenon of **quantum tunneling** may also be examined by HTTM in the same direction: before measurement determines the outcome, the wave function possesses an amplitude extending into and through the classically forbidden region. However, tunneling should not be identified with "escaping from time"; this is a distinct interpretation of HTTM that would need to be developed into quantitative predictions if it is to be distinguished from standard quantum mechanics.

Observation→Interface→HTTM→Temporalization t2

CHAPTER VIII: THE BIG BANG, BLACK HOLES, CONFORMAL CYCLIC COSMOLOGY (CCC), AND CREATION WITHIN THE FRAMEWORK OF THE TIMELESS DOMAIN — TLD

Summary

The studies of Stephen Hawking, James Hartle, Jacob Bekenstein, and Roger Penrose have opened new perspectives on the origin, evolution, and ultimate fate of the universe. The Hawking–Hartle ***No-Boundary*** *model proposes that the universe does not necessarily have a sharply defined starting point in time, while Penrose's* ***Conformal Cyclic Cosmology (CCC)*** *proposes that the universe passes through multiple cycles (aeons), with each cycle succeeding the previous one through conformal geometric structures without requiring a contraction (Big Crunch) followed by another explosion.*

Research on black holes has also profoundly changed the understanding of modern physics. From initially being regarded merely as geometric solutions of general relativity, black holes gradually came to be understood as physical systems possessing entropy and temperature. The work of Jacob Bekenstein and Stephen Hawking raised a central problem in modern physics: what happens to information when matter falls into a black hole? This is the ***black hole information paradox****, one of the questions that still has no complete solution.*

According to the TLD model proposed in this book, a black hole is not necessarily a place where information is destroyed, but may instead represent the final stage of the process of material actualization. As entropy approaches its maximum, the structure of the Interface Domain becomes unstable; space collapses, time nearly ceases to manifest, and matter gradually loses its form. In this interpretation, a black hole may be regarded as a recycling mechanism that returns information from the material world to a deeper informational level of the Timeless Domain (TLD), rather than as the absolute endpoint of information.

This view also suggests a new interpretation of Penrose's CCC model. If TLD is the timeless foundation of all phenomena, then each aeon may be regarded as a different cycle of actualization of the same informational foundation. The end of an aeon is not the destruction of reality, but the return of information to the TLD before a new cycle of actualization is initiated.

From this perspective, the question ***"When did the universe begin?"*** *needs to be understood differently. If time appears only after information from TLD is actualized within 4D spacetime, then the concept of "when" has meaning only within a particular cosmic cycle. For the Timeless Domain itself, there is no beginning or ending point in the sense of physical time. Therefore, the absolute origin cannot be measured in years, seconds, or by the age of the universe.*

This view bears a philosophical similarity to the Buddhist concept of ***beginninglessness (anādi)****. The Buddha taught that the cycle of birth and death is beginningless and that no absolute starting point of samsara can be found. In the TLD model, this is interpreted to mean that time is only an attribute of the phenomenal world, whereas the timeless informational foundation is not governed by the concepts of before and after. Therefore, every question concerning a "first moment" has meaning only within the domain of spacetime and does not apply to the ultimate foundation of reality.*

TLD (timeless)➔ → Big Bang / beginning of an aeon → evolution of the universe→ increase in entropy→ black hole→ information returns to TLD→ a new aeon.

I. Introduction to the Big Bang in Modern Physics

In modern physics, the Big Bang is generally regarded as the very early stage of the universe, when all observable spacetime was in an extremely hot, extremely dense state and was expanding very rapidly. However, it should be made clear from the outset: the Big Bang should not be understood as an "explosion" of matter from a small point into an empty surrounding space. More precisely, the Big Bang is the expansion of space itself. There is no "outside" into which the universe exploded. Space, time, matter, and radiation emerged together within a common process. NASA also describes the history of the universe in this way: after the inflationary period, energy was transformed into matter and light, after which the universe continued to cool and the light elements were formed.

In Einstein's equation, when time is traced backward toward t→0t \to 0, the scale factor a(t)→0a(t)\to 0. At that point, density, temperature, and spacetime curvature tend toward infinity. This is what is called a **singularity**. But a singularity is not necessarily an ultimate physical reality. It may simply indicate that general relativity has been pushed beyond the limits of its applicability.

The important point is this: the Big Bang describes the earliest boundary of physical time that present-day science can trace backward, but it does not answer the deeper questions: Why did time emerge? Why did spacetime have to be activated? And before physical time existed, can one meaningfully speak of "before" or "after"?

It is precisely here that the **Timeless Domain — TLD** model may be introduced as an ontological hypothesis.

According to the standard cosmological model, the Big Bang is supported by three major groups of evidence: the expansion of the universe, the cosmic microwave background radiation, and the relative abundances of light elements such as hydrogen, helium, and lithium. The Big Bang was not a mechanical explosion within space, but the expansion of the entirety of spacetime from an extremely hot and dense state. The Harvard-Smithsonian Center for Astrophysics also emphasizes that during the Big Bang era there was no empty region of space outside; all space was relatively uniformly filled with matter and energy.

But this does not necessarily mean that the universe truly began from an "infinitely dense material point." Rather, it may simply mean that when the Planck regime is reached, Einstein's continuous spacetime model is no longer sufficient. A deeper theory is required: quantum gravity, string theory, M-theory, loop quantum gravity, or another ontological model.

II. Hawking's View: No Boundary, No Absolute Starting Point

Stephen Hawking, together with James Hartle, proposed the **"no-boundary proposal."** According to this view, the universe does not necessarily have a sharply defined beginning resembling an edge or boundary in time. Instead, time in the extremely early universe may behave like a spatial dimension; therefore, the question "What was before the Big Bang?" may become inappropriate. The Hartle–Hawking model uses the path integral of quantum mechanics and the concept of imaginary time to describe a geometry that is finite but without boundary, analogous to the surface of the Earth, which is finite but has no edge.

This point is very important for TLD. Hawking did not say that there was "nothing" and then suddenly there was an explosion of matter. He suggested that in the primordial regime, the very concept of linear time loses its meaning. Therefore, one cannot ask "What was before it?" in the ordinary sense.

TLD can accept this point but go further. According to TLD, before physical time existed, there was not absolute nothingness, but rather a timeless domain containing primordial structural information. There is no before–after, no event, and no physical process there. When information needs to be unfolded, time is activated as an organizing mechanism.

Thus, TLD does not reject the Big Bang. TLD only states that the Big Bang is not the ultimate origin. It is the threshold at which physical time emerges within the present aeon.

III. Black Holes, Entropy, and Information: A Philosophical–Physical Perspective

A **Black Hole** forms when matter and energy are compressed to such an extent that gravity overcomes all opposing pressure. In general relativity, when a region of space is compressed below its

Schwarzschild radius, an event horizon appears. From the outside, no signal from within can escape according to classical physics. From this perspective, a black hole is merely a geometrical system of spacetime described by Einstein's equations. However, in the 1970s, the research of Jacob Bekenstein and Stephen Hawking completely changed this understanding.

Jacob Bekenstein (1947–2015) raised a seemingly simple question: what happens to information when a book, a hard drive, or even an entire library is thrown into a black hole? If all information truly disappears, the laws of thermodynamics would be violated because the entropy of the universe would decrease. From this argument, he proposed that a black hole must possess entropy, meaning that it must preserve information in some form.

Surprisingly, the entropy of a black hole is not proportional to its internal volume but to the area of its event horizon:

S=kBc3A4GℏS=\frac{k_Bc^3A}{4G\hbar}

where AA is the area of the event horizon. This result suggests that all information associated with the black hole appears to be encoded on its surface rather than within its volume. This is one of the foundations that led to the famous **holographic principle**: a three-dimensional world may be described by information stored on a two-dimensional boundary.

The next discovery was even more surprising. Hawking demonstrated that black holes are not completely black. Because of quantum fluctuations near the event horizon, a black hole emits thermal radiation and therefore has a temperature:

TH=ℏc38πGMkBT_H=\frac{\hbar c^3}{8\pi GMk_B}

The temperature is inversely proportional to mm.

Thus, an object once regarded as purely geometrical possesses the properties of a thermodynamic system: entropy and temperature. This is one of the most beautiful discoveries of modern physics—the unexpected meeting of three concepts that had appeared entirely different: **gravity, entropy (disorder or lack of information), and temperature**. This convergence occurs in one of the strangest objects in the universe: the black hole. According to the conventional view, gravity is created by matter, entropy represents disorder, and temperature arises from collisions among particles.

Black Holes as the Endpoint of a Cycle of the Material World

If entropy SS continues to increase:

Sentropy→Smax,S_{\text{entropy}}\rightarrow S_{\max},

then the TLD Interface becomes unstable.

A black hole may be regarded as a place where: space collapses, time freezes, and matter loses its form.

From this perspective, a black hole is not the destruction of information.

It is the process of returning information from the material domain to a deeper foundational level.

TLD→ (S,T,H,m)→ Death→TLD	or	TLD→ (S,T,H,Matter)→BH→TLD
Human		Universe

IV. Penrose's CCC

Roger Penrose proposed **Conformal Cyclic Cosmology — CCC**. According to this model, the universe passes through multiple aeons. Each aeon begins in a state resembling a Big Bang and ends in the extremely distant future. By then, the stars have gone out, massive matter gradually loses its role, black holes evaporate through Hawking radiation, and the universe becomes extremely dilute. Penrose proposes that when mass no longer plays a decisive role, conformal geometry allows the infinite future of the preceding aeon to connect with the Big Bang of the following aeon.

CCC is not a Big Crunch followed by another explosion in the ordinary sense. It does not require the entire universe to collapse into a point and then explode again. Instead, it connects the extremely distant future of one aeon with the beginning of the next through conformal structure.

This point is highly compatible with TLD if interpreted as follows:

- CCC describes the geometric cycle of spacetime.
- TLD describes a deeper ontological level where time can withdraw and be reactivated.

- CCC is a physical–geometrical model.
- TLD is an informational–ontological model.

Therefore, TLD can regard CCC as the physical trace of a deeper process: deviated information requires time in order to be unfolded, corrected, dissolved, and then reorganized.

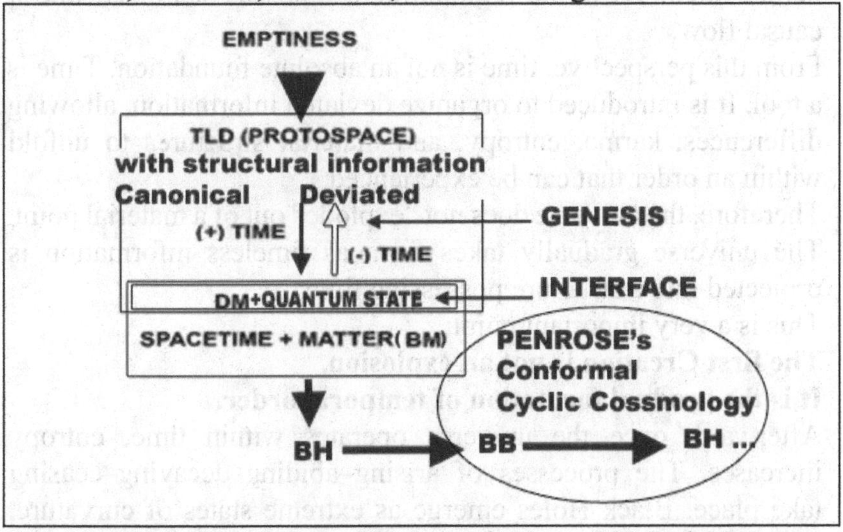

BH: Black Hole, BB: Big Bang, BM: Baryonic Matter, DM: Dark Matter

V. The First Creation Is Not the Big Bang

In TLD, Two Levels Must Be Distinguished: Ontological Creation

This is the point at which timeless information begins to require unfolding into a temporal order. This is not the physical Big Bang. It is the activation of the principle of time.

At this level, there is not yet matter in the 4D sense, no photons, no atoms, no stars, and no galaxies. There is only the transition from the timeless domain toward the possibility of time.

This May Be Called: The First Creation

Once the principle of time has been activated, physical aeons may emerge repeatedly. Each Big Bang in CCC or in a cyclic universe is only a new manifestation of spacetime.

VI. TLD, 11D, and the Formation of 4D

According to the TLD model, the deepest reality is not 4D. 4D is only the phenomenal domain in which time, space, matter, and causality have already become stabilized. The process may be envisioned as follows:

Emptiness → TLD → Interface → 4D Spacetime

TLD has no time. The Interface is the region where time begins to be activated. 4D is the region where time has stabilized into a causal flow.

From this perspective, time is not an absolute foundation. Time is a tool. It is introduced to organize deviated information, allowing differences, karma, entropy, and material structures to unfold within an order that can be experienced.

Therefore, the universe does not "explode" out of a material point. The universe gradually takes form as timeless information is projected into a structure possessing time.

This is a very important point:

The first Creation is not an explosion.

It is the gradual formation of temporal order.

Afterward, once the universe operates within time, entropy increases. The processes of arising–abiding–decaying–ceasing take place. Black Holes emerge as extreme states of curvature, entropy, and the withdrawal of time. In the extremely distant future, in the spirit of **Conformal Cyclic Cosmology/CCC**, a new aeon may emerge.

VII. Why Can Human Beings Not Know When the Universe Began?

If time appears only after TLD is projected into 4D, then the question "When did the universe begin?" has meaning only within the present aeon. We may ask: How long has our observable universe been expanding? But we cannot ask: When did absolute reality begin? Because "when" already presupposes the existence of time.

If there is a timeless domain, it has no starting point measured by a clock. Therefore, the absolute origin cannot be measured in years, seconds, cycles, or the age of the universe.

This is close to the Buddhist concept of **beginninglessness**. In the *Saṃyutta Nikāya*, the Buddha states that samsara is beginningless and that no beginning can be found for beings wandering through birth and death. The deeper meaning is that once we have entered the flow of time and dependent conditions, we cannot use that same temporal flow to locate an absolute beginning.

VIII. Relation to Genesis

In the Book of Genesis, the creation narrative should not simply be understood as a physical record of the Big Bang. It resembles a symbolic structure describing the manifestation of order from a formless foundation.

"In the beginning God created the heavens and the earth."

"The Spirit of God moved upon the face of the waters."

"Let there be light."

The light here is not necessarily physical photons. It may be understood as the first differentiation, the first opening of order, the emergence of the capacity for awareness. Light is mentioned before the Sun, Moon, and stars. This suggests that light in Genesis carries an ontological meaning more than a physical one.

Within TLD, this may be interpreted as:

- **The first day:** primordial differentiation of information.
- **The second day:** polarization, stratification, and formation of domains.
- **From the third day onward:** the gradual densification of information into material structures.
- **The sixth day:** the appearance of living beings possessing knowledge and ego.

Adam and Eve eating the forbidden fruit may be understood as a fall into discriminative knowledge. Before this was a state close to timelessness, without birth and death. Afterward came labor, suffering, reproduction, and death—that is, the temporal domain. Thus, Genesis does not necessarily conflict with TLD. It may be read as a symbolic description of the transition from timelessness into time.

IX. Relation to Buddhism

Buddhism does not focus on a primordial Big Bang. Buddhist scriptures commonly describe worlds through cycles of **formation–abiding–decay–emptiness**. In the *Aggañña Sutta*, the world undergoes dissolution and then formation again after extremely long periods. There are periods without the Sun, Moon, day, night, months, or years. This is very close to the idea that physical time has not always existed in the way human beings experience it.

Buddhism also speaks of innumerable worlds, innumerable Buddha-realms, and innumerable kalpas. In a broad sense, this corresponds to the idea of multiple universes or multiple aeons.

Within TLD, one may say:

- **Nirvana / H-TLD:** no arising or ceasing, no entropy, no need for time.
- **L-TLD:** informational deviation remains, with potential still requiring unfolding.
- **4D:** a temporal environment in which karma, entropy, differences, and correction can operate.
- **Black Hole:** a place where time withdraws from physical structure.
- **Big Bang:** a place where time is reactivated.

Therefore, the universe arises and ceases repeatedly, while the timeless foundation neither arises nor ceases.

Diagram 3.

The concept of cyclic BB and the Multiverse after the initial Creation; therefore, the beginning of BB should be regarded as unknowable.

X. Synthesis: Big Bang, Black Hole, and TLD

The model may be summarized as follows:

Timeless TLD

In this model, the Big Bang is not the first Creation. The Big Bang is a physical event within time. The first Creation is the emergence of the very possibility of time.

Therefore, it may be stated:

The Big Bang is not the absolute origin of reality, but the phenomenological threshold at which physical time begins within an aeon. Creation is deeper than the Big Bang, because Creation is the ontological activation of time from the timeless domain. A Black Hole is the reverse process: a state in which entropy, curvature, and informational deviation reach their extreme, causing time to withdraw from the 4D structure. Penrose's CCC may be regarded as a geometrical model for the succession of aeons, while TLD provides the ontological foundation for why time can be activated, withdrawn, and reactivated. Therefore, the observable universe may have a finite age, while foundational reality is beginningless in the Buddhist **sense: an absolute starting point cannot be traced by means of physical time itself.**

Conclusion

From the standpoint of modern physics alone, the Big Bang is a highly successful model, but it does not completely resolve the problem of origins. Hawking softens the starting point through the no-boundary model. Penrose extends cosmic history into multiple aeons through CCC. Bekenstein and Hawking show that the Black Hole is where entropy, information, and gravity meet.

TLD can bring these intuitions together within a broader framework:

- Hawking shows that linear time may lose its meaning at the origin.
- Penrose shows that the Big Bang may not occur only once.
- Black Holes show that time, entropy, and information are deeply interconnected.
- Buddhism describes the universe as beginningless and undergoing cycles of formation–abiding–decay–emptiness.
- Genesis describes Creation as the manifestation of order from an invisible foundation.

TLD proposes that time is not the ultimate foundation. Time is a mechanism introduced to unfold and correct deviated information. Therefore, Creation does not begin with an explosion, but with the emergence of the possibility of time. The Big Bang is only the physical imprint of a deeper process.

CHAPTER IX: LANGUAGE, NUMBERS, AND GEOMETRY

Summary

This chapter proposes an expanded perspective on mathematics, physics, and philosophy based on the Timeless Domain (TLD). According to this model, before the emergence of matter, energy, and spacetime, there existed a domain of primordial structural information in which relationships, order, and symmetries existed in a timeless form. Spacetime, matter, and the laws of physics are regarded as the results of the temporalization and realization of these informational structures.

From this perspective, language, arithmetic, and geometry are not the ultimate foundations of reality, but rather sequential manifestations of deeper informational structures. Language is the temporalized form of information; mathematics is the language that describes part of the order of TLD; and physics is the manifestation of mathematical relationships in the phenomenal world.

The chapter also analyzes the roles of postulates, paradoxes, Gödel's Incompleteness Theorems, and Hilbert space (a space for mathematical equations). A postulate is the starting point of every theoretical system; a paradox is a sign that reveals the limits of a system; Gödel's theorem indicates that no sufficiently powerful formal system can prove its own completeness and consistency; while Hilbert space suggests that quantum reality is broader than what is observed within spacetime. According to the TLD model, these results all suggest that reality may contain a deeper foundational layer beyond spacetime.

The TLD model proposes that many paradoxes of modern physics, such as the Big Bang, black holes, quantum measurement, and quantum entanglement, do not necessarily reflect contradictions in nature, but may instead indicate the limitations of models based solely on 4D spacetime. When reality is expanded to include the timeless domain, these paradoxes may be understood as manifestations of the same informational structure viewed from two different levels of reality.

From these analyses, the chapter proposes several foundational postulates of TLD: time is not an ultimate reality but a phenomenon or manifestation of the mechanism for organizing information; primordial structural information exists prior to matter; the 4D world is the manifestation of information through the Interface Domain; spacetime cannot fully explain its own origin; entropy reflects the dispersion of informational structure; and under extreme conditions such as black holes, information may withdraw from material form and return to a deeper foundational level.

The chapter also examines the question of why mathematics is capable of describing the universe with such precision. From Galileo, Wigner, and Penrose to Tegmark, various perspectives have been proposed to explain the relationship between mathematics and nature. According to the TLD model, mathematics neither creates reality nor is identical with reality; rather, both mathematics and the material world reflect the same timeless informational foundation. Because they share a common origin, mathematics becomes an appropriate language for describing the universe.

Finally, the chapter emphasizes that truth is not identical with proof. Proof is a product of dualistic thought within spacetime, whereas truth belongs to the intrinsic structure of reality. Mathematics is a bridge between human intelligence and the order of TLD, but it is not the ultimate foundational level. From this perspective, the process of cognition may be summarized as follows:

Emptiness → Informational Structure → Mathematics → Physical Laws → Matter → Life → Knowledge → Recognition of Truth.

Thus, the ultimate purpose of the TLD model is not to replace mathematics or modern physics, but to propose a deeper ontological foundation in which mathematics, physics, consciousness, and the phenomenal world are understood as different manifestations of the same timeless informational foundation.

I. Cognition and Language

In TLD, information exists in a complete, unfragmented, and non-dual state. However, in the 4D world—when time and duality appear—information must be encoded sequentially through language, symbols, and writing. Therefore, writing systems are not the origin of information, but rather the partial unfolding of information that already exists in TLD.

Thus, both geometry and language can be regarded as projections of a deeper informational order. Their development over time is not the creation of new structures, but the gradual manifestation of an underlying timeless structure.

The above observation is not entirely consistent with the materialist view that *"consciousness itself is its immediate reality,"* in which "reality" is understood primarily as material life, and language is the direct form of expression of consciousness—a process in which the subject speaks to itself.

In contrast, according to the argument of this work, consciousness is not the direct product of matter, but the manifestation of information originating from the Timeless Domain (TLD). Information in TLD exists in a timeless, complete, and unfragmented state; when it enters the temporal 4D world, it is gradually "unfolded" into cognitive structures, from which consciousness is formed.

From this perspective, matter and language are not the foundations that generate consciousness, but intermediate means through which information already existing in TLD is manifested. Language, therefore, is not consciousness itself, but the sequential encoding of timeless informational content into the temporal and dualistic structure of the physical world.

Language Is the Temporalized Form of Information

In the work *Tiếng Việt huyền diệu (2004)*, Nguyễn XQ presented persuasive arguments that language and writing evolve and develop over time in the course of human development. This is consistent with the view that language is the sequential manifestation, within a temporal world, of deeper informational structures that are themselves independent of time. Language and writing evolve over time throughout human development. This does not contradict the TLD perspective; on the contrary, it suggests a profound correspondence: language in the 4D world is the sequential, historical, and cultural manifestation of deeper informational structures that are independent of time.

In TLD, there is no speech, writing, or grammar in the ordinary sense, because these elements arise only when there are time, bodies, brains, communities, and communication. However, TLD may be hypothesized to contain pre-linguistic structures: order, relationships, distinctions, symmetry, and the capacity to generate meaning. When these structures are projected into the biological and historical domains, they gradually manifest as sounds, symbols, words, syntax, and writing.

Therefore, language is not the ultimate foundation of reality. **Language is the temporalized form of information.** It allows human beings to express, communicate, and preserve a small portion of the deeper order within TLD.

II. Arithmetic, the Timeless Domain, and the Limits of Proof

In the physical 4D world, time plays the role of organizing and "unfolding" these pre-existing relationships into sequential cognition. Therefore, mathematical reasoning and proof may be understood as temporal processes through which timeless structures are approached and reconstructed.

Arithmetic may be a manifestation of deeper informational relationships that are independent of time. The natural numbers 1,2,3,…,n1,2,3,\ldots,n do not themselves "progress"; they exist as a structure possessing an intrinsic timeless order. In this sense, arithmetic may be understood as originating from a timeless informational foundation—the Timeless Domain (TLD).

From this perspective, unsolved problems—such as the **Goldbach conjecture**—may be reconsidered. Although many specific cases, such as 4=2+24=2+2, 6=3+36=3+3, and 10=5+5=3+710=5+5=3+7, satisfy the conjecture, a general proof has not yet been obtained. Rather than asserting that it cannot be proved, this may be regarded as a manifestation of the limitations of formal systems based on logic. *(Note: Goldbach's conjecture states that every even number is the sum of two prime numbers. A prime number is divisible only by 1 and itself.)*

This perspective is compatible with Gödel's incompleteness theorem, which shows that in every sufficiently powerful mathematical system, there exist true propositions that cannot be proved within that system itself. From the TLD perspective, this suggests that mathematical truth may extend beyond what can be sequentially deduced.

Therefore, reality may be understood as including both spacetime and a timeless informational domain. Mathematics and science, when restricted solely to a temporal framework, may inadvertently overlook this deeper level.

Euclidean Geometry, Axioms, and the Manifestation of Temporal Structure

Within the framework of the Timeless Domain (TLD), geometrical relationships may be understood as manifestations of the internal consistency of information rather than as results of time-dependent processes. From this perspective, Euclid's first axioms—particularly Postulates 1 and 2, concerning drawing a straight line between two points and extending a straight line indefinitely—may be regarded as fundamental timeless relationships. They do not require motion, duration, or transformation; rather, they reflect the immediate consistency of informational structure, compatible with a timeless domain.

From the TLD perspective, this suggests that some geometrical truths arise directly from timeless informational relationships, whereas others reflect the gradual manifestation of these relationships when interpreted within a framework involving distinction, extension, and increasing complexity—characteristics associated with the emergence of time.

A. POSTULATES, PARADOXES, GÖDEL'S THEOREM, AND HILBERT SPACE WITHIN THE TLD FRAMEWORK

1. What Is a Postulate?

Every system of thought begins with certain propositions that are not proved from within that system. These are called **postulates** or **axioms**.

In Euclidean geometry, we accept certain axioms concerning points, lines, and planes. In Newtonian physics, we accept absolute space and time. In Einstein's relativity, we accept that spacetime can curve and that the speed of light is an invariant limit. In quantum mechanics, we accept that a quantum state is represented by a vector in Hilbert space.

Thus, a postulate is not an "absolute truth." A postulate is the starting point of a descriptive system.

The deeper a system becomes, the fewer but more powerful its postulates may be. However, no system can be entirely free of postulates. If we wish to prove one postulate, we must appeal to another, deeper postulate. Therefore, reason always requires a foundation that it cannot itself completely prove.

This is very important for TLD. If 4D physics postulates that spacetime is fundamental, then TLD proposes a different postulate: spacetime is not the ultimate foundation; the deeper foundation is a timeless informational domain.

2. What Is a Paradox?

A paradox appears when a system of reasoning leads to a result that contradicts intuition or appears self-contradictory.

A classical example is the liar paradox:

"This sentence is false."

If the sentence is true, then it is false. If it is false, then it is true.

A paradox is not merely an error of language. It shows that a system has encountered a limit. When a system permits excessively strong self-reference, it may generate an unstable loop.

There are also paradoxes in physics:

- the black hole information paradox;
- the quantum measurement paradox;
- the problem of time in quantum gravity;
- the Big Bang singularity;
- the Black Hole singularity.

These paradoxes do not necessarily mean that reality is irrational. They often mean that our current language, mathematics, or postulates are not yet deep enough.

In other words:

Paradox=a sign of the limits of a system→ lack of a dêlė foundation level

In TLD, paradoxes arise when we use 4D language to describe what lies prior to 4D. If we ask, "What was before the Big Bang?", we have already used the concept of "before"—that is, time—to ask about a state in which time did not yet exist. Thus, the question itself already contains a paradox.

3. Gödel's Incompleteness Theorems

Kurt Gödel proved two famous incompleteness theorems. According to the Stanford Encyclopedia of Philosophy, the first theorem states that in any consistent formal system sufficiently powerful to contain basic arithmetic, there will be propositions that can neither be proved nor disproved within that system itself. The second theorem states that such a system cannot prove its own consistency from within itself.

In brief:

If a system is sufficiently powerful and consistent, then it is incomplete. Such á in cáe ò Quantum Mechanic Intêpretation

And:

A system cannot completely prove its own consistency.

This is not a failure of mathematics. It is a profound discovery: **truth is broader than formal proof.**

A proposition may be true but unprovable within the system under consideration. To prove it, one must step outside that system and use a broader system.

This has profound philosophical significance:

No sufficiently powerful system within duality can be absolutely self-contained.

The Meaning of Gödel for TLD

Gödel shows that mathematics possesses an intrinsic limitation: a system cannot complete itself entirely from within.

TLD may interpret this analogously at the ontological level:

4D cannot completely explain 4D from within itself.

If we remain solely within spacetime, we can describe phenomena occurring within spacetime. But it becomes difficult to answer: Why does spacetime exist? Why are there laws? Why is there mathematical order? Why does time emerge?

These questions resemble the "Gödel propositions" of 4D physics. They are meaningful, but cannot be completely resolved using language internal to 4D.

Therefore, TLD plays the role of a "broader system." It does not deny 4D, nor does it deny physics. It only proposes that 4D may be a subsystem, while the deeper foundation is the timeless domain of structural information.

4. What Is Hilbert Space?

Hilbert space is the mathematical space used to represent quantum states. In quantum mechanics, the state of a physical system is represented by a vector in Hilbert space, which is a complex vector space equipped with an inner product.

Put more simply: in classical physics, an object has a position and velocity in 3D space. But in quantum mechanics, a particle does not necessarily possess a single definite position before measurement. It has a possible state containing multiple superposed possibilities.

Therefore, Hilbert space is the space of quantum possibilities, not ordinary physical space.

Hilbert Space and the Domain of Possibility

Hilbert space shows that modern physics has moved beyond classical material intuition. Quantum reality is not simply a collection of tiny particles flying through space. It is a structure of possibilities in which a physical state is a combination of multiple possibilities.

This may be expressed in the language of TLD as follows:

- Hilbert space is the mathematical domain of quantum possibilities.
- 4D is the phenomenal domain of manifested outcomes.
- TLD is the deeper domain of timeless structural information.

Hilbert space is not TLD. It remains a mathematical structure within quantum physics. But it provides a very useful intermediate image: what is observable in 4D is only a small part of a broader space of possibilities.

5. Paradox as a Gateway to a Deeper Level

Throughout the history of thought, paradoxes have often not been dead ends. They have been openings.

Zeno's paradoxes forced humanity to think more deeply about motion and continuity. The paradoxes concerning light led Einstein toward relativity. The black-body radiation problem led to quantum theory. The Black Hole information paradox stimulated holography. Gödel's theorem shattered the dream of a completely self-contained formal system.

Therefore, the paradoxes associated with the Big Bang, Black Holes, quantum measurement, and consciousness should not necessarily be regarded as minor errors requiring repair. They may be traces indicating that reality is deeper than our current models.

In TLD, paradoxes arise because we are viewing a timeless level through the lens of time. Just as the shadow of a three-dimensional object on a two-dimensional plane may appear paradoxical to a 2D being, traces of TLD in 4D may appear paradoxical to purely 4D physical reasoning.

Paradox, Duality, and the Non-Dual Origin of Postulates

Paradox arises only when reality is viewed through a dualistic structure. Within the dualistic domain, thought must distinguish true from false, existence from non-existence, subject from object, before from after, and cause from effect. This polarization makes reasoning possible, but at the same time creates the conditions for paradox.

The liar paradox is one example. The statement "I am lying" becomes paradoxical only because we force it into one of two values: true or false. If it is true, then it is false; if it is false, then it is true. The paradox does not lie in ultimate reality, but in the dualistic structure of language and logic.

Similarly, quantum paradoxes such as superposition, EPR entanglement, or "tunneling" appear paradoxical only when we use the dualistic intuition of the classical world to understand a deeper level of reality. In the ordinary 4D world, an object must be here or there, an event must occur before or after another, and systems must be independent or not independent. But in quantum mechanics, states may consist of combinations of possibilities, and systems may be related in ways that do not conform to classical spacetime intuition.

Therefore, when TLD is included within reality as a whole, paradox need not represent a genuine contradiction. It may merely indicate that dualistic reasoning is encountering a deeper non-dual level. What the 4D domain calls a paradox may simply be the distorted shadow of a higher order when projected into spacetime.

Likewise, a postulate cannot be completely proved through dualistic reasoning. Every system of reasoning requires a starting point. That starting point is not generated by proof within the system but is accepted as its foundation. Thus, a postulate may be understood as a trace of non-duality within the dualistic domain: it is something upon which reasoning must depend but which reasoning cannot itself create.

From this perspective, unproven propositions in arithmetic, such as Goldbach's conjecture, may be regarded as signs that mathematics is not completely enclosed within a formal system of proof. This is not sufficient to prove that TLD exists, but it suggests that arithmetic may touch a level of order deeper than the present proving capacity of dualistic reason.

In other words, what cannot be proved within the dualistic domain is not necessarily meaningless. It may be a gateway leading back toward a deeper non-dual foundation. **In TLD, truth does not begin with proof; proof begins only after truth has been projected into the domain of distinction.**

III. Some TLD Postulates

Based on the preceding analyses, TLD may be presented through several postulates:

Postulate 1: The Timelessness Postulate

Ultimate reality does not begin with time. Time is a mechanism activated to organize information.

Postulate 2: The Structural Information Postulate

Prior to matter, energy, and spacetime, there exists primordial structural information.

Postulate 3: The Manifestation Postulate

The 4D world is the finite manifestation of timeless information through the Interface.

Postulate 4: The 4D Incompleteness Postulate

4D spacetime cannot completely explain its own origin.

Postulate 5: The Entropy–Deviation Postulate

Entropy in 4D may be understood as the physical manifestation of informational deviation, dispersion, and loss of order.

Postulate 6: The Regression Postulate

When the 4D structure can no longer maintain temporal order, information withdraws from physical form, as in the Black Hole limit.

Summary

A postulate is the starting point of a system. A paradox is a sign that the system has encountered its limits. Gödel shows that no sufficiently powerful formal system can complete itself entirely from within. Hilbert space shows that quantum reality is broader than the outcomes observed in 4D. TLD takes these lessons and extends them into ontology.

If Gödel is correct in mathematics, then we should be humble in assuming that 4D physics can explain the entirety of reality. If Hilbert space is correct in quantum mechanics, then what is observable represents only part of a broader space of possibilities. If TLD is correct, then matter, time, and the universe are not the ultimate foundation, but conditional manifestations of timeless information.

Therefore, the ultimate question is not merely:

When did the universe begin?

But rather:

Why was time activated so that the universe could appear?

And the answer proposed by TLD is:

Time is introduced as a mechanism for organizing, unfolding, and correcting deviated information. 4D is the stage upon which this process takes place. TLD is the timeless foundation behind the stage.

In TLD, truth does not begin with proof; proof begins only when truth has been projected into the domain of distinction.

IV. MATHEMATICS AND REALITY

From Galileo to the Timeless Domain

A. Why Can Mathematics Describe the Universe?

One of the strangest features of modern science is that the universe appears to be written in the language of mathematics. Every fundamental law of physics can be expressed through mathematical equations. From the motion of planets, the propagation of light, and atomic structure to the evolution of the entire universe, all can be described by mathematical expressions with remarkable precision.

What is noteworthy is that mathematics is not first discovered from nature and only then invented. In many cases, mathematical structures are developed in a completely abstract manner, only to be found decades later to describe the physical world precisely.

This leads to a very profound question:

Why does mathematics correspond so well with nature?

B. Galileo: Nature Is Written in the Language of Mathematics

Galileo was the first to state very clearly:

"The book of nature is written in the language of mathematics."

According to Galileo, without understanding geometry and mathematics, human beings cannot read the book of the universe.

This was a major turning point. He did not regard mathematics merely as a computational tool. Mathematics itself is the language of nature.

However, Galileo did not explain **why** nature possesses such a language.

C. Wigner: The Unreasonable Effectiveness of Mathematics

In 1960, Eugene Wigner published his famous paper:

The Unreasonable Effectiveness of Mathematics in the Natural Sciences.

He was astonished by the phenomenon that mathematics is often developed entirely independently of physics, yet many years later turns out to describe nature accurately.

For example:

- Riemannian geometry became the foundation of general relativity;

- complex numbers became fundamental to quantum mechanics;
- Lie groups became the language of particle physics.

Wigner called this “a wonderful gift which we neither understand nor deserve.” Even today, physics has not fully explained why mathematics is so “effective.”

D. Penrose: Mathematics Exists Objectively

Roger Penrose proposed a view close to Platonic philosophy. According to him, the mathematical world exists objectively and independently of human beings. Humans do not create mathematics. Humans merely discover it.

Penrose proposed three worlds:

- the physical world;
- the mental world;
- the mathematical world.

These three worlds are deeply interconnected, although their relationship has not yet been fully explained. This perspective explains why mathematics has universality, but it still does not answer the question: Why does the physical world obey the mathematical world?

E. Tegmark: The Universe Is Mathematics

Max Tegmark proposed an even stronger and more radical hypothesis. According to him:

The universe is not merely described by mathematics; the universe itself is a mathematical structure.

According to the Mathematical Universe hypothesis, every physical reality is simply a self-consistent mathematical structure. Therefore, there is no need to ask:

“Why does mathematics describe nature?”

because:

“Nature itself is mathematics.”

However, one question remains unanswered:

If the universe is mathematics, where does mathematics come from?

TLD: Mathematics Reflects the Structure of the Timeless Domain

The Timeless Domain proposes another possibility. According to this hypothesis, mathematics does not create reality; rather, mathematics reflects part of the structure of the timeless informational domain.

In other words, TLD is deeper than mathematics. In TLD, arithmetic, geometry, and equations do not exist in the sense understood by human beings. There exist only intrinsic relationships, structures, symmetries, and informational order. The mathematics constructed by human beings is merely an abstract language reflecting a very small portion of these structures.

This may be represented as:

TLD→Informational Structure→Mathematical Relations→Physical Laws→4D Universe

Thus, mathematics is not the cause that creates the world. It is like a musical score, while TLD is like the thought of the composer. The score reflects the thought, but does not generate the thought.

Why Is Mathematics Effective?

From the TLD perspective, mathematics is effective not because the universe *is* mathematics, but because both mathematics and the universe reflect the same deeper informational foundation.

Therefore, the meeting of mathematics and physics is not miraculous, but a natural consequence of their common origin. This also explains why purely mathematical structures may, many years later, become the language of physics. Mathematicians do not create the laws. They gradually discover structures that already exist at a deeper level of reality.

G. TLD and the Limits of Mathematics

If Gödel is correct, then no mathematical system can be absolutely complete from within itself. This suggests that mathematics, like physics, may be only a manifestation of a deeper foundation.

According to TLD, that foundation is the timeless domain. Within that domain, there are not yet proofs, theorems, or axioms. These appear only when informational structure is projected into the dualistic domain of thought.

Therefore, mathematics is not the ultimate starting point, but a bridge between human thought and the structure of the Timeless Domain.

H. WHY IS TRUTH ALWAYS SIMPLER THAN MATHEMATICS?

A Perspective from the Timeless Domain

"Mathematics is the language of science, but truth does not necessarily have to be written in any language."

1. Truth and Its Expression

Throughout the history of science, mathematics has always been regarded as the most precise language of thought. Every important physical law can be expressed through equations. For this reason, many people have concluded that reality itself is mathematics.

However, it is necessary to distinguish between **truth** and **the expression of truth**.

A map is not the territory. A musical score is not the composer. An equation is not reality.

Mathematics helps describe reality, but mathematics itself is not reality.

2. Truth Always Precedes Proof; Truth in TLD Does Not Require Proof Because There Is No Irrationality in TLD

In mathematics, a theorem is accepted only after it has been proved. But proof does not create truth. For example:

The sum of the angles of a Euclidean triangle was 180∘180^\circ before Euclid proved it. The law of gravity was true before Newton discovered it. Gravitational waves existed for billions of years before humans observed them.

Thus, proof does not create truth.

Proof only helps the human intellect approach truth.

3. Gödel and the Limits of Proof

Gödel's incompleteness theorem shows that, in every sufficiently powerful axiomatic system, there will always be true propositions that cannot be proved from within that system.

This has profound philosophical significance. If there are truths that exceed the capacity of proof, then truth is greater than mathematics.

Or, more precisely:

Truth is not identical with proof.

Mathematics is only a window through which we look toward truth. That truth originates from a level deeper than mathematics—that is, TLD.

4. Why Is Truth Simple?

In nature, fundamental laws are often very simple.

Einstein's equation contains only a few symbols. Schrödinger's equation is also very short. Maxwell's equations can be written on a single page. Yet from these simple equations, the richness of the entire universe emerges.

This suggests that:

The foundation is always simpler than its manifestation.

5. Mathematics Becomes Increasingly Complex

By contrast, as human beings attempt to describe the entirety of reality, mathematics becomes increasingly complex:

Hilbert space.

Riemannian geometry.

Lie group theory.

String theory.

M-theory.

Why does it become increasingly vast and complex?

Not because truth itself is complex, but because finite intelligence must employ many layers of symbols to describe a reality that it does not directly perceive.

6. TLD: Truth Precedes Mathematics

According to the Timeless Domain, equations do not exist in the timeless domain. Arithmetic does not exist there. Nor do geometry, tensors, or Hilbert space.

What exists are intrinsic relationships, order, symmetry, and informational structure. These structures do not need to be written as equations in order to exist.

Mathematics appears only when the dualistic intellect attempts to describe these structures.
Therefore, mathematics does not generate TLD.
TLD itself is also not mathematics.
Mathematics is merely a mirror reflecting part of TLD.

7. Why Does Mathematics Have Limits?

If mathematics reflected the entirety of reality, then every truth would have to be provable. But Gödel shows that this is not the case.

If mathematics were identical with reality, then quantum paradoxes should not exist. Yet they still do. If mathematics were the ultimate foundation, we would not need to ask, "Why is mathematics correct?" But this very question astonished Wigner. This suggests that mathematics is not the ultimate level.

8. Truth in TLD

In the Timeless Domain, truth does not require proof.

It is simply the intrinsic structure of reality. Only when reality is projected into the temporal domain, the dualistic domain, and discriminative intelligence do logic, definitions, axioms, proofs, and theorems appear.

It may be said that proof is a product of thought, whereas truth is a property of reality.

The two are not identical.

9. A New Principle

From the preceding analyses, a TLD principle may be formulated:

The Primordial Principle of Truth

Truth→Mathematics→Physics→Matter→Life→ Insight / Reflection→Truth

(without accumulating additional equations) — Timeless Domain.

Conclusion

Galileo argued that nature is written in the language of mathematics.

Wigner was astonished that mathematics describes nature with such precision. Penrose proposed that mathematics exists objectively. Tegmark went further by identifying the universe with mathematical structure.

The Timeless Domain proposes a different perspective. Mathematics does not create reality. Mathematics is also not identical with reality. Mathematics is a mirror that directly reflects part of the structure of the timeless informational domain.

If this hypothesis is correct, then the "unreasonable effectiveness" of mathematics is no longer mysterious. It is simply because mathematics and the universe both reflect a common foundation: the **Timeless Domain**.

Within TLD, the concept of "proving the existence of TLD" has not yet been defined, because proof requires language, logic, a knowing subject, and an object that is known. These elements arise only after reality has manifested within the domain of distinction and time. Therefore, TLD cannot be directly proved in the ordinary sense. It can only be indirectly recognized through the consistency, order, and unifying capacity of the structures for which it serves as the foundation. /+

www.ingramcontent.com/pod-product-compliance
Lightning Source LLC
LaVergne TN
LVHW041207061126
851857LV00018B/1972